D0072177

DATE DUE

MAR 1 1 2002			

Demco, Inc. 38-293

MANIPULATION
ROBOTS

MANIPULATION
ROBOTS
Dynamics, Control, and Optimization

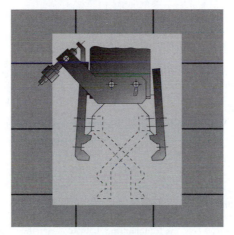

Felix L. Chernousko
Nikolai N. Bolotnik
Valery G. Gradetsky

Institute for Problems in Mechanics
Russian Academy of Sciences
Moscow, Russia

CRC Press
Boca Raton   Ann Arbor   London   Tokyo

Mitchell Memorial Library
Mississippi State University

Library of Congress Cataloging-in-Publication Data

Chernous 'ko, F. L.
    [Manipuliatsionnye roboty. English]
    Manipulation robots : dynamics, control, and optimization / by
Felix L. Chernousko, Nikolai Bolotnik, and Valery G. Gradetsky.
      p. cm.
    Translation of: Manipuliatsionnye roboty.
    Includes bibliographical references and index.
    ISBN 0-8493-4457-3
    1. Robots--Dynamics.   I. Bolotnik, N. N. (Nikolaĭ Nikolaevich)
II. Gradetskiĭ, V. G. (Valeriĭ Georgievich) III. Title.
TJ211.4.C4713 1993
629.8'92--dc20

                            93-18977
                                CIP

This book contains information obtained from authentic and highly regarded sources. Reprinted material is quoted with permission, and sources are indicated. A wide variety of references are listed. Reasonable efforts have been made to publish reliable data and information, but the author and the publisher cannot assume responsibility for the validity of all materials or for the consequences of their use.

Neither this book nor any part may be reproduced or transmitted in any form or by any means, electronic or mechanical, including photocopying, microfilming, and recording, or by any information storage or retrieval system, without prior permission in writing from the publisher.

CRC Press, Inc.'s consent does not extend to copying for general distribution, for promotion, for creating new works, or for resale. Specific permission must be obtained in writing from CRC Press for such copying.

Direct all inquiries to CRC Press, Inc., 2000 Corporate Blvd., N.W., Boca Raton, Florida 33431.

© 1994 by CRC Press, Inc.

No claim to original U.S. Government works
International Standard Book Number 0-8493-4457-3
Library of Congress Card Number 93-18977
Printed in the United States of America 1 2 3 4 5 6 7 8 9 0
Printed on acid-free paper

THE AUTHORS

Felix L. Chernousko, Ph.D., D.Sc., is a well-known scientist in the fields of control theory, mechanics, and applied mathematics.

Professor Chernousko graduated from Moscow Physico-Technical Institute in 1961 and received his Candidate of Sciences (Ph.D.) degree in 1964. His Ph.D. dissertation was devoted to the dynamics of satellites. In 1969, Professor Chernousko received his Doctor of Sciences degree for his dissertation concerning the dynamics of rigid bodies containing fluid masses.

From 1964 to 1968, Professor Chernousko worked in the Computing Center at the U.S.S.R. Academy of Sciences, where he made an essential contribution to numerical methods of optimal control. Since 1968, he has worked at the Russian Academy of Sciences as a Department Head in the Institute for Problems in Mechanics. He is also a Professor at the Moscow Physico-Technical Institute.

Professor Chernousko has made important contributions to control theory and methods. He has also carried out essential research in the area of robotics. In 1980 he was awarded a U.S.S.R. State Prize for his research of mechanics. In 1992 he was elected as a Full Member of the Russian Academy of Sciences (Academician).

Professor Chernousko has published more than 200 research papers and nine monographs. He has also presented a number of invited lectures at international conferences. He is a member of the editorial boards of several Russian and international scientific journals, including the *Journal of Applied Mathematics and Mechanics, Optimal Control Applications and Methods,* and the *Journal of Optimization Theory and Applications.*

Nikolai N. Bolotnik, Ph.D., D.Sc., is an experienced scientist in the fields of theoretical mechanics, mechanical engineering, and control theory.

He graduated from Moscow Physico-Technical Institute in 1973, with a major in control theory. In 1978, he received his Candidate of Sciences (Ph.D.) degree for defending his thesis dealing with the optimization of shock isolation systems. In 1992, he received his Doctor of Sciences degree for his dissertation devoted to optimal control of manipulation robots.

Professor Bolotnik has worked as a researcher at the Russian Academy of Science's Institute for Problems in Mechanics since 1973. He is also an Assistant Professor at Moscow Physico-Technical Institute.

Professor Bolotnik has made a significant contribution to the theory of optimal shock and vibration isolation of engineering systems, and to the development of methods of optimal control for robots. He is the author or joint author of more than 100 scientific publications, including two monographs. He is a member of the editorial board of the Russian Academy of Science's *Journal of Engineering Cybernetics.*

Valery G. Gradetsky, Ph.D., D.Sc., has a broad reputation as a scientist in the areas of robotics, automation technology, and sensory systems.

In 1958, he received his M.S. degree in mechanical engineering from Bauman Moscow High Technical School, now Moscow State Technical University. In 1964, he received his Candidate of Sciences (Ph.D.) degree from the same school. His dissertation dealt with space technology and the development of control fluidic elements for hazardous environments. In 1984, he received his Doctor of Sciences degree for his dissertation on the theory and development of sensory systems for robots subject to external disturbances acting on the robots grippers.

From 1958 to 1975, Professor Gradetsky worked at the Institute for Control Sciences, U.S.S.R. Academy of Sciences. He dealt with the development of extremal controllers and the first pneumatic robots. From 1975 to 1981, he was Department Head at the Research Institute of Introscopy working on the problems of sensor fusion in inspection automation technology and robotic systems. Since 1981, he has worked as the Laboratory Head at the Institute for Problems in Mechanics, Russian Academy of Sciences. He is also a Professor of Robotics and Automation.

Professor Gradetsky has made a great contribution to the development of manipulation-intelligent robots and mobile wall-climbing robots. He has been awarded several medals for achievement in research and development of mechatronic automation systems and robots.

Professor Gradetsky is the author or joint author of over 200 research papers and patents, as well as four books. He has also presented many lectures at international conferences. He is a member of the Russian Academy of Science's Scientific Council of Scientific and Technology Association, and the Scientific Council of the Technological Academy.

TABLE OF CONTENTS

Chapter 8

Applications of Robots 229

INTRODUCTION

Manipulation robots find many industrial applications and are used for various technological operations: assembly, welding, cutting, painting, etc. Robots enhance the labor productivity in industry and deliver a man from tiresome, monotonous, or hazardous work. Moreover, robots make many operations better than a man does, and they provide higher accuracy and repeatability. In many fields, high technological standards are hardly attainable without robots.

Apart from industry, manipulation robots are used in extreme environments, e.g., in space or on the ocean bottom. They can work in chemical-active media, in the locality contaminated with radioactive materials, at low and high temperatures, etc. Manipulation robots can also be helpful in everyday life. For example, they can assist in housekeeping and serving handicapped people. Thus, manipulation robots are prospective machines whose application area is widening.

From the engineering point of view, manipulation robots are complex, versatile devices which contain a mechanical structure, a sensory system, and an automatic control system. Theoretical fundamentals of robotics rely on the results of research in mechanics, electronics, automatic control, and other sciences. For example, mechanical properties of robots influence such basic performance characteristics as positioning accuracy and productivity. It is very important also to know the possibilities of the robot control and to be able to make the best (optimal) use of these possibilities.

In this book, we systematically expound some fundamental issues of modern robotics, primarily concerning dynamics and optimal control of manipulation robots. We focus mostly on the topics related to the author's research work at the Institute for Problems in Mechanics of the Russian Academy of Sciences. The well-known material that can be found in textbooks and monographs on robotics (e.g., kinematics and dynamics of manipulators with rigid links) is presented in this book briefly. Despite the brevity, the presentation of basic kinematics and dynamics is sufficient to comprehend the main material without referring to other books.

The main part of the book is original and covers the following issues: the dynamics of manipulation robots with elastic links and/or joints; the optimal control of manipulators; and the sensor systems of robots. These aspects are of great practical importance, because they are directly connected with the accuracy, productivity, and versatility of robotic systems.

The material of the book is arranged in eight chapters.

Chapter 1 is introductory. Here, we give a basic knowledge of the structure of manipulation robots, describe their kinematics, information and control systems, and different types of actuators.

In Chapter 2, we briefly expound on basic kinematics and dynamics of manipulators with rigid links.

Chapters 3 through 7 are devoted to the dynamics of robots with elastic parts. Chapter 3 presents the general theory of mechanical systems with elastic parts of high stiffness. The approach developed here is used in the subsequent chapters for the analysis of the dynamics of flexible manipulators.

In Chapter 4, we study the behavior of manipulation robots with elastic links.

Chapter 5 is devoted to the experimental analysis of elastic properties of industrial robots. The technique for static and dynamic testing, as well as the results of the experiments for concrete manipulation robots, are presented here.

In Chapter 6, we study the dynamics of some industrial robots with elastic joints by using the method of Chapter 3. The theoretical results are compared with the experimental data from Chapter 5. In Chapter 6, we also describe a software package for computer simulation of the dynamics of flexible manipulation robots.

Chapter 7 deals with the optimal control of manipulation robots. For different models of robots, time-optimal and suboptimal control laws (both open-loop and feedback) are constructed by using analytical and numerical techniques. It is shown that optimal control makes it possible to considerably reduce the operational time of industrial robots.

In Chapter 8, we discuss some new applications of robots. We describe here robotic systems for measurements and inspection, special equipment for them (sensors and sensory grippers), as well as wall-climbing robots capable of performing different technological operations.

We hope that our book will be helpful to researchers and engineers dealing with problems of modern robotics, as well as to students of mechanical engineering, electrical engineering, and automatic control.

The investigations reflected in the book were carried out at the Institute for Problems in Mechanics. We are thankful to our colleagues L. D. Akulenko, V. V. Avetisyan, A. I. Grudev, A. A. Gukasyan, A. A. Kaplunov, S. A. Mikhailov, G. U. Pirumov, M. Yu. Rachkov, N. N. Rogov, and V. B. Veshnikov for their valuable contributions to the results covered in the book. We thank R. P. Soldatova for her great help in preparing the manuscript.

Chapter 1

STRUCTURE AND BASIC COMPONENTS OF MANIPULATION ROBOTS

1.1. MANIPULATION ROBOTS AND THEIR KINEMATIC STRUCTURE

1.1.1. BASIC INFORMATION

The robots, new mechatronic controlled machines, began to be widely used in industry and other branches of human activity in the late 1960s. Since then, robots have become more and more involved in automatic manufacturing, in operations in hazardous environments, and even in everyday life. They can perform work that requires high accuracy and dexterity and is difficult, dangerous, or unpleasant for man. For instance, robots can manipulate with radioactive isotopes, work under water, in chemically aggressive media, or under high temperature, and they assist in serving disabled people and hospital patients. Modern robots not only produce "physical" work according to specified programs, but also develop an intelligent behavior. They can gather information about the environment and its own state, process this information, and make reasonable decisions on how to accomplish the prescribed tasks. Unlike traditional automatic systems, robots permit changing production and service operations flexibly within a wide range.

Depending on their purposes and characteristics of implemented tasks, robots are divided into manipulation, locomotion, inspection, measuring, information, control, etc. At present, a new family of robots, called personal robots, is under development. They are aimed at helping man in everyday life. The majority of robots combine operations of different types, for example, manipulation and measurement.

The most widely used are manipulation industrial robots, which carry out different operations that a man would perform with his hands. To some extent, manipulation robots imitate human arm motions.

According to internationally adopted terminology, a manipulation industrial robot is an automatically controlled, reprogrammable, multipurpose manipulative machine with several degrees of freedom, which may be either fixed in place or mobile, for use in industrial automatic applications.

Let us explain some terms used in this definition. Reprogrammable means that planned motions of the robot can be altered by changing the control code only, without any physical alterations. Multipurpose means that the robot can be adapted to different applications, perhaps by means of physical alterations. By physical alterations we mean any structural changes as well as a replacement of one tool with another. Changes of floppy disks, cassettes, read-only-memory (ROM) chips, etc., are not considered physical alterations.

3

According to the definition of the Industrial Federation of Robotics, industrial robots are divided into several types, depending on their control systems. They distinguish:

1.   Nonservo, point-to-point (PTP), or sequence-control robots
2.   Servo, pose-to-pose, continuous path (playback), and computer numerical control robots
3.   Adaptive, or intelligent, robots (these robots include many sensors in their feedback loops)
4.   Computer-aided teleoperators, or teleoperating robots

From the technological and design points of view, the manipulation robot is a mechatronic controlled system including one or several manipulators, a control unit, drives, transducer and sensor systems, grippers, and other technological equipment. Manipulation robots are intended for carrying out the technological or maintenance operations associated with motion of workpieces or tools in accordance with a prescribed task.

In modern manufacture, technological robotic installations are widely used for such operations as welding (arc and spot), assembly, painting, cutting, drilling, inspection, grinding, etc., depending on the tools set in the robot gripper. Robot motions resemble those of a man working with different technological equipment (machine tools, conveyors, laser, plasma or nuclear installations, etc.).

A manipulator is a mechatronic system with a programmed or manual remote control, which brings workpieces and tools to the desired place inside the working (service, operation) zone and implements the prescribed operations. The working zone is the room part accessible for the gripper (with a tool or a workpiece).

The general structure of an industrial manipulation robot is given by the block diagram in Figure 1.1.1. The robot includes the manipulator (1), the control system (2), the drives (3), the sensor system (4), transducers (5), technological equipment and tools (6), and interface devices (7), connecting different units of the robot. The control system can be connected with different terminals (8), manual control (9), and external memory units (10).

The sensor (or information) system of the robot is usually included in the feedback loop. This system processes the primary environment information received from the sensors of physical (geometrical, mechanical, etc.) parameters, forms the generalized data, and transmits them into the control system. By using a monitor, it is possible to organize the interaction between the sensor system and an operator. The control system also envisages the dialog of the operator and the robot by means of a terminal unit (a keyboard, a control panel, a printer, a graphical device, a console, etc.).

The intelligent behavior of the robot depends on algorithms used for data processing and decision making.

The manipulator shown in Figure 1.1.1 is placed on the base (11), which can be either fixed or mounted on a mobile vehicle (12). The manipulator

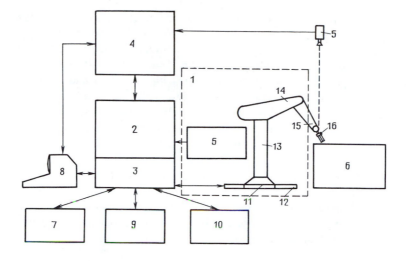

FIGURE 1.1.1.

comprises the arm consisting of the links (13 to 15) and the gripper (16). The base, the robot body, the arm, and the gripper form the mechanical system of the robot. A body handled in the robot gripper (e.g., workpieces, intermediate products, tools, special instruments, etc.) is called a manipulation object.

Note that the concepts of manipulation robot and manipulator do not coincide totally. A manipulation robot may include one or several manipulators. On the other hand, manipulators can be operated independently. However, the terms automatically controlled manipulator, automatic manipulator, and manipulation robot are often used as synonyms. Sometimes, industrial robot is defined as reprogrammable automatic manipulator.

With respect to the control methods, manipulation robots are divided into manually controlled (biomechanical), automatic, and interactive ones. Biomechanical robots follow the motion of an operator's arms. Automatic robots receive information through their sensor systems and are controlled automatically using feedback. Interactive or supervision robots combine the automatic and manual control modes. Here the operator can interfere in the automatic process and modify the robotic operations.

There are different approaches to classification of industrial robots. The basic classification criteria are

1.   The functional purpose
2.   The application field determined by the types of technological operations
3.   The kinematic structure
4.   Operation and maintenance characteristics

The functional purpose is usually distinguished by technological robots, transport-and-service robots, and universal robots. Technological robots are involved in the manufacturing process as technological machines; they directly implement basic manufacturing operations. Transport-and-service robots automatically perform auxiliary operations (picking up tools and setting them in the technological machines, transporting workpieces and manufactured articles, etc.). They provide the communication between machine tools and technological robots, on the one hand, and conveyors, storages, and warehouses on the other. Universal robots can carry out both basic and auxiliary operations.

Depending on the fields of application, industrial robots are divided into welding, assembly, casting, punching, machining, and painting robots and robots for automatic measurement and inspection, galvanization, etc. Some robots can perform several operations simultaneously, for example, operations of transport and automatic inspection.

1.1.2 KINEMATIC STRUCTURES OF MANIPULATORS

From the viewpoint of kinematics, a manipulator is a system of links connected by joints. The manipulator is to be designed so that a manipulation object could acquire desired positions and orientations, within the limits implied by the technological purpose of the robot. The manipulator kinematic structure is determined by the types of the joints connecting the links. Usually, the links are connected consecutively and form an open kinematic chain. The first link is connected to a base and the last one carries a gripper, or an end-effector. Sometimes, manipulator kinematic structures may include loops.

The most widely used are revolute and prismatic (rectilinear) joints. The links connected by a revolute joint can rotate relative to each other about a common axis. The prismatic joint admits relative translations of the links along the straight line. More sophisticated joints can also be used in manipulators, for example, the screw joint which combines the prismatic and revolute joints in one unit, or the spherical joint.

Figure 1.1.2 gives conventional graphic presentations for some types of joints. The diagrams a, b, and c show the prismatic, spherical, and screw joints, respectively. The diagrams d, e, f, and g correspond to revolute joints of different configurations.

The kinematic structure of a manipulator determines the generalized coordinates which are the most convenient for describing the motion of the manipulator. If the motion of the end-effector of the manipulator results from the superposition of rectilinear motions of three links in three mutually orthogonal directions, Cartesian coordinates are suitable. Cylindrical coordinates are used for the manipulator whose end-effector moves due to changing the length of the arm, its rotation in the basic coordinate plane, and the translation in the direction perpendicular to this plane. Spherical coordinates are the most convenient for the robot whose end-effector moves by changing the length of the arm and its rotations about two mutually orthogonal axes.

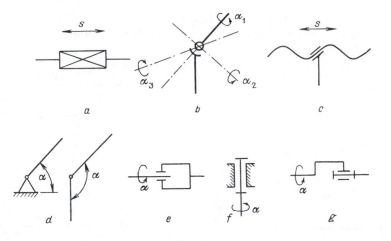

FIGURE 1.1.2.

To describe the kinematics of multilink manipulators with revolute joints (articulated manipulators), the angular system of generalized coordinates is used. As a rule, articulated manipulators are controlled by the drives placed at the joints. The minimal number of generalized coordinates sufficient to specify any possible configuration of the manipulator is called the number of degrees of freedom of this manipulator. Usually, the number of degrees of freedom coincides with the number of independently controlled drives.

The shape of the working (service) zone of the manipulator depends on its kinematic structure and the constraints on relative displacements of the links.

Kinematic diagrams for some types of manipulators are presented in Table 1.1.1. The table also gives projections of the working zones of the manipulators onto two mutually orthogonal planes. Row (1) corresponds to the Cartesian coordinate manipulator whose links 1, 2, and 3 can move in three mutually orthogonal directions X, Y, and Z. Row (2) shows the cylindrical coordinate robot. Links 1 and 3 can move in mutually orthogonal directions R and Z, while link 2 rotates about the Z axis (φ is the rotation angle). Row (3) gives the diagrams for the spherical coordinate manipulator. The motion of the gripper results from the translation of link 1 in the direction R and two rotations of links 2 and 3 (by the angles φ and ψ) about mutually orthogonal axes. The last row (4) corresponds to the angular coordinate robot whose links 1, 2, and 3 can rotate with respect to each other (by the angles φ_1, φ_2, and φ_3).

The manipulator motions and the corresponding degrees of freedom are usually divided into transport, orientation, and auxiliary motions. Transport motions provide the arrival of the gripper at the given place of the working zone. These motions are performed by the links of the manipulator arm. Orientation motions are implemented by the links of the gripper and give the

TABLE 1.1.1.
Basic Kinematic Diagrams of Manipulators

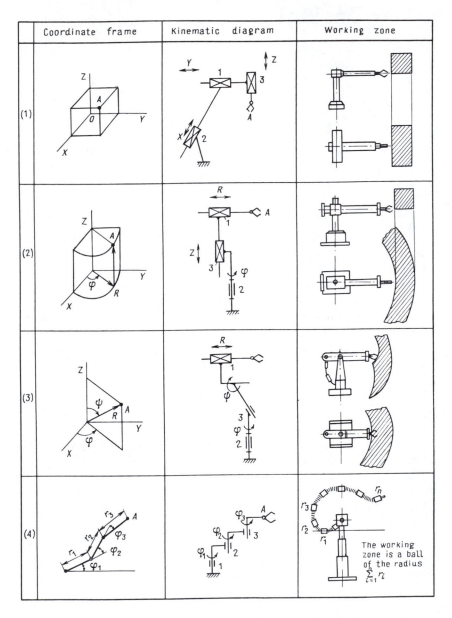

	Coordinate frame	Kinematic diagram	Working zone
(1)			
(2)			
(3)			
(4)			The working zone is a ball of the radius $\sum\limits_{i=1}^{n} r_i$

desired orientation to the end-effector. Auxiliary motions include the motion of the robot base, grasping motions of the gripper jaws, etc.

Different mechanisms of grippers are shown in Figure 1.1.3. Grasping can result from rotations of links 1 and 2 by angles α_1 and α_2 about common

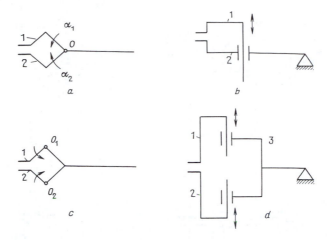

FIGURE 1.1.3.

axis O (a); from the translation of link 1 with respect to fixed link 2 (b); from rotations of links 1 and 2 about the axes of joints O_1 and O_2 (c); from translations of links 1 and 2 with respect to fixed body 3 (d).

1.1.3. ROBOTS OF DIFFERENT DESIGN

Industrial robots are divided into several types, depending on their design. A robot whose base is mounted on the shop floor is called a floor robot. A robot with the base mounted on a trolley which can move along a transverse bar placed at some altitude is called a gantry robot. Gantry robots save working area and provide definite advantages in serving the manufacturing machines, their maintenance, and repair. If the transverse bar is long enough, the same robot can serve several machine tools.

Floor robots with horizontal telescopic arms are widely used for pick-and-place operations; for example, they serve punching presses. Such robots perform limited vertical displacements, have small response time, and are equipped with a simple pose-to-pose control system. As examples of these robots, we may mention Versa-Tran® (U.S.); Ritm and Universal (Russia); Autohand, Motoman®, and Seiko® Instruments RT 5000 (Japan), etc.

Robot Ritm (Figure 1.1.4a) performs transport motions in Cartesian reference frame according to the kinematic diagram shown in Figure 1.1.4b (for one arm). The robot body and the arm can move in X, Y, and Z directions inside the working zone. In addition, the gripper can rotate by the angle α.

The robot of the Versa-Tran® type with two arms is shown in Figure 1.1.5a. The kinematics of transport motions for this robot is presented in Figure 1.1.5b. The corresponding degrees of freedom are the rotation of the body by the angle φ, its vertical translation in Z direction, and independent extension of the arms in the directions X_1 and X_2. In addition, the arms can rotate by the angles α_1 and α_2.

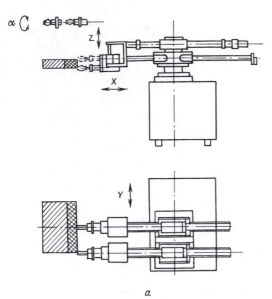
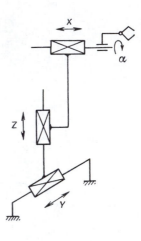

a   b

FIGURE 1.1.4.

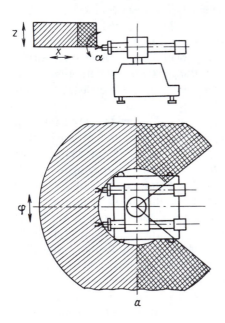
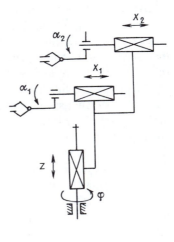

a   b

FIGURE 1.1.5.

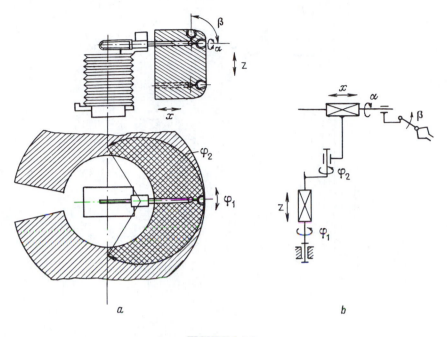

FIGURE 1.1.6.

Robot Universal-5 (Figure 1.1.6a) is a cylindrical coordinate robot. Its transport motions (Figure 1.1.6b) result from the body rotation by the angle φ_1, the arm rotation by the angle φ_2, the vertical displacement in the direction Z, and the horizontal displacement of the arm in the direction X. The gripper can rotate by the angle α and swing by the angle β.

Floor robots with horizontal telescopic arms mounted on a vertically sliding carriage are usually cylindrical or Cartesian coordinate robots. As examples of such robots we can point out Fanuc® M (Japan), AMF® Versa-Tran® (U.S.), Modular Robot (France), LM40C4701 (Russia), etc.

The latter robot has a modular structure, its different versions being assembled by proper connecting of the standard modules as shown in Figure 1.1.7. On the mobile base driving the robot body in the direction X, the column is mounted which can rotate by the angle φ. Along the vertical guide on the column, the sliding carriage carrying the arm can move in the direction Z. The arm can extend by the distance r and rotate by the angle α. The robot is supplied with sensors connected with the drives of the basic motions (column rotation, carriage sliding, and arm displacement). These sensors are included in the feedback circuit.

Arc welding robot Shin Meiwa PW 752 (Japan) performs translations in X, Y, and Z directions of the Cartesian reference frame (Figure 1.1.8). The gas burner, in addition, can rotate about a vertical axis by the angle φ and swing by the angle α.

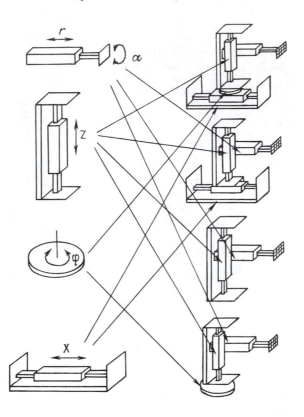

FIGURE 1.1.7.

Cylindrical and spherical coordinate robots with swinging telescopic arms are often used for such technological operations as automatic arc and spot welding, polishing, grinding, etc. Some examples are robots Prab, Stanford Arm, Kawasaki®-Unimate (U.S.), Mitsubishi® (Japan), Universal-15 and Universal-60 (Russia), etc. Robots Unimate and Universal-15 (Figure 1.1.9a) are spherical coordinate robots. The arm of the robot Universal-15 has an additional degree of freedom, the translation parallel to the swing axis. The robot is equipped with a hydraulic drive and a hydraulic amplifier connected with a DC electric motor. The kinematic structure of the robot (see Figure 1.1.9b) provides decoupled rotation and swing motions of the gripper.

Articulated manipulators with multilink arms moving in a vertical plane have comparatively large working zones. As examples, we mention robots Puma®, Cincinnati Milacron® (U.S.), IRb-60 ASEA® (Sweden), Toshiba®, Mitsubishi®, Mazak® (Japan), Kuka® (Germany), and RPM-25 (Russia).

Robots of the Puma® type have three transport degrees of freedom (body, upper arm, and forearm rotations) and two orientation degrees of freedom (rotations of the wrist about two axes); see Figure 1.1.10. Load-carrying capacities of such robots are not large (1 to 10 kg); they have a small mass

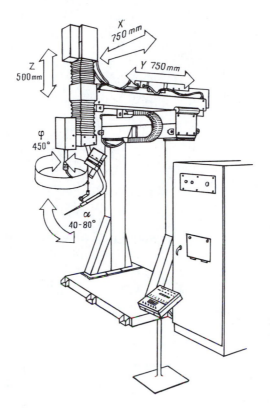

FIGURE 1.1.8.

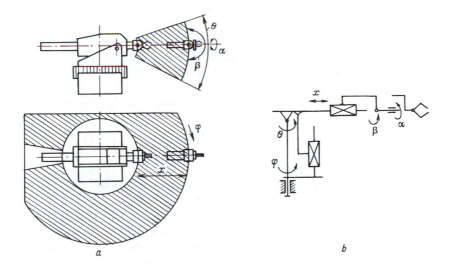

FIGURE 1.1.9.

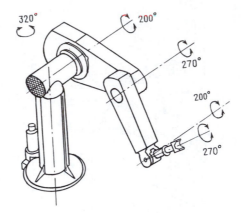

FIGURE 1.1.10.

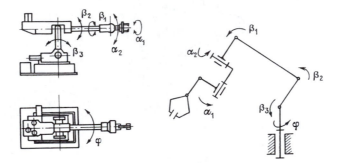

FIGURE 1.1.11.

(up to 120 kg) and high positioning accuracy (0.05 to 0.2 mm). These robots are suitable for precise operations such as assembly, automatic inspection of materials and manufactured articles, precise setting of workpieces in machine tools, welding, etc.

Robot RPM-25 (Figure 1.1.11), driven by electromechanical direct current (DC) actuators, is universal and can be supplied with different control systems. It can be used for arc welding of thick sheet workpieces, for service of forging of metalworking equipment, for laser technology, etc.

Multilink SCARA type robots (Figure 1.1.12a) were developed especially for assembly operations in the microelectronic industry. The basic motions of these robots occur in a horizontal plane. This family of robots includes Skilam (Japan), Sony® (IBM® and Sony®, U.S.-Japan), Nam Robo (Japan), etc. These robots use the impulse control motors placed at the joints. The kinematics of SCARA type robots admit rotations in the shoulder and elbow joints (by the angles α_1 and α_2), rotation of the gripper by the angle α_3 (all rotations about vertical axes), and the vertical displacement of the gripper in

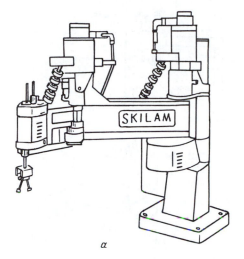

a

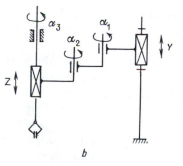

b

FIGURE 1.1.12.

the direction Z. In addition, the manually controlled displacement in direction Y is provided, with subsequent fixation of the set position.

A typical example of the gantry robot used in large-scale manufacturing is robot CM40F2.80.01 (Russia), shown in Figure 1.1.13. The manipulator of this robot has two arms mounted on a carriage which can move along a monorail. The robot kinematics admits two swinging motions of each arm, in the shoulder and elbow joints, thus enabling the grippers to travel inside the working zone whose area is 30 m². This robot is mostly aimed at setting the workpieces in machine tools with horizontal spindle axes. The robot has an electrohydraulic step drive. Gantry robots are usually supplied with a set of removable grippers and tools. They can be built-in inside machine tools, automatic lines, or flexible manufacturing systems.

Many modern manipulation robots can change tools (drills, gas or plasma burners, screwdrivers, etc.) in their grippers automatically. This increases the

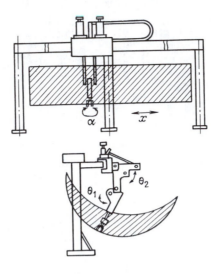

FIGURE 1.1.13.

flexibility of robots in accomplishing manufacturing tasks. Grippers of different types (vacuum, magnetic, mechanical, etc.) can be used, depending on manipulation objects.

Thus, robots must conform to their role in the manufacturing process and working conditions. This explains a wide variety of existing kinematic structures and designs of robots and their operational parameters (the number of degrees of freedom, load-carrying capacity, driving torques, grasping forces, positioning accuracy, shape and area of the working zone, etc.). Note that the positioning accuracy is a significant parameter for all types of robots, especially for those aimed at precise operations such as assembly or welding.

1.2. BASIC TYPES OF DRIVES

1.2.1. GENERAL REQUIREMENTS

Requirements imposed on robot drives depend on the technological process to be robotized. In some cases, for example, it is sufficient to implement operations by rigid PTP programs, while others require feedback controls taking into account position, velocity, driving torques, or external forces.

Depending on power supply, the drives can be classified as electric, pneumatic, hydraulic (electrohydraulic), etc. Robot drives should have high power characteristics and small dimensions. The latter is especially important for drives built-in inside manipulator joints. Robot drives must be suitable for automatic controllers, should have quick response time, and be able to implement sophisticated control laws, including optimal and adaptive controls.

The ideal diagram of motion for a typical robot drive is shown in Figure 1.2.1a,b. It consists of acceleration ($\ddot{x} > 0$), constant velocity ($\ddot{x} = 0$), and

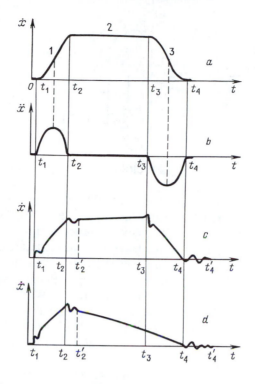

FIGURE 1.2.1.

deceleration ($\ddot{x} < 0$) parts marked by numbers 1, 2, and 3, respectively. Here, x is a coordinate, e.g., a joint angle. Due to finite response time of drives, oscillations may appear when the velocity changes abruptly. In particular, it takes place when the acceleration-deceleration curve has a trapezoidal (Figure 1.2.1,c) or triangular (Figure 1.2.1d) shape.

It is important that the drive should work smoothly, without jerks and shocks, especially for robots carrying out precise technological operations. It can be achieved by the proper design of the drive mechanisms as well as by using appropriate controllers. As a rule, controllers for robot drives have feedback circuits.

Robot drives should have hard load characteristics, i.e., their maximal loading torque must depend weakly on the angular velocity of the output shaft of the drive within a sufficiently wide range. In addition, the robot drives should satisfy the requirements of stability, reliability, and longevity.

1.2.2. ELECTRIC DRIVES

Electric drives are widely used in industrial robots. These drives are compact, easy to control, have minimal response time, and do not require complex equipment, hydraulic or pneumatic pumps, pipelines, hoses, etc.

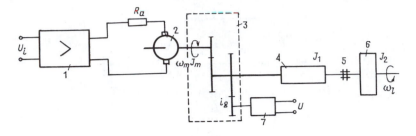

FIGURE 1.2.2.

Due to their compactness, electric drives offer a designer more freedom in placing them on the robot. In particular, electric drives can be built-in inside manipulator joints, which is convenient for articulated robots.

One of the prospective types of drives is a stepper electric drive, especially when it is used with a computer control system. This drive consists of a stepper motor and a reduction gear. Both open-loop (programmable) and feedback controllers are used for the stepper motor. When the stepper motor is controlled by a rigid program, the number of control impulses is proportional to the angular or linear displacement of the corresponding manipulator link, and the frequency of these impulses determines the angular velocity of the motor shaft.

At present, DC motors are most widespread in manipulation robot drives. DC motors have hard load characteristics and the moments of inertia of their rotors are comparatively small. The electric drive with a DC motor may include power amplifiers, reduction gears, and sensors for internal information in the feedback circuits. Depending on the technological tasks and working conditions of the robot, these sensors can measure position, velocity, driving torques, current in the armature circuit, etc. Sensors are also used in the robot units which restrict displacements of the links, set them automatically in the initial (home) positions, or fix the links in the desired positions. Reduction gears for robot drives must have small backlashes to transmit rotations without jerks and provide high positioning accuracy for the robot. These requirements are met by the wave gear train consisting of the generator of deformation waves and flexible and rigid gears.

The structure of the drive with a DC motor is shown in Figure 1.2.2. Here, the numbers indicate a DC amplifier (1), a motor (2), a gear train (3), a manipulator link (4), a connection unit (5), a load (6), and a tachogenerator (7). The letters denote the input voltage (U_i); the moments of inertia of the motor (J_m), manipulator link (J_1), and the external load (J_2); the electric resistance of the armature circuit (R_a); the output voltage of the tachogenerator (U); angular velocities of the motor shaft (ω_m) and output shaft (ω_l); and the gear ratio (i_g).

Examples of robots equipped with electric drives are Puma®, Skilam, Universal, Cincinnati Milacron®, etc. (see Section 1.1.3).

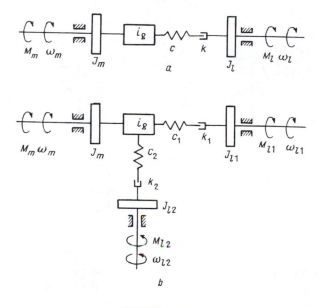

FIGURE 1.2.3.

In mathematical models of manipulation robots (see Chapter 2), electric drives are often considered as linear dynamical systems. It should be noted, however, that different nonlinear imperfections such as dry friction, dead zones, backlashes, etc., are always present in real drives. These nonlinearities influence performance capabilities of the robot, in particular its accuracy. Sometimes, the influence can be significant and should be taken into account in the mathematical model. Parallel with the nonlinearities, the performance capabilities of robots, especially their accuracy and productivity, can also be affected by elastic compliance of links, joints, and parts of driving mechanisms.

Figure 1.2.3 shows the models of mechanical parts of drives taking into account the elasticity and backlashes at output shafts of gear trains. Figures 1.2.3a and 1.2.3b correspond to the schemes with one and two output shafts, respectively.

The letters denote the torque created by the electric motor (M_m); the angular velocities of the motor shaft (ω_m) and output shafts $(\omega_l, \omega_{l1},$ and $\omega_{l2})$; the inertia moments of the motor (J_m) and external leads $(J_l, J_{l1},$ and $J_{l2})$; the gear ratio (i_g); the stiffness coefficients $(c, c_1,$ and $c_2)$ and the backlash parameters $(k, k_1,$ and $k_2)$ of the gear trains; and the load torques $(M_l, M_{l1},$ and $M_{l2})$.

Individual parts of the driving mechanism (motor, gear trains, etc.) are usually assembled as a single unit called a power module.

In robot drives, the angular velocity of robots is often controlled using electronic (thyristor) transducers. The rotation velocity can be continuously

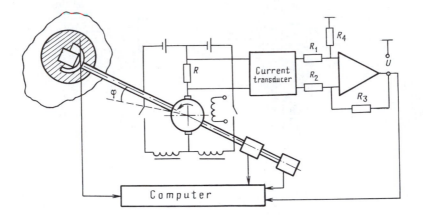

FIGURE 1.2.4.

adjusted within wide limits by changing the voltage applied to the armature circuit, keeping the driving torque constant. The latter is achieved by changing the excitation magnetic flux.

It is important to investigate the dependence between voltages and currents in motor circuits, on the one hand, and the generated driving torques/forces on the other. This dependence is used in computed-torque control algorithms. It is also important to develop procedures for separating in the current response the component proportional to the forces/torques of interaction of robots with technological equipment. The proportionality can be used for measuring external forces.

Automatization of technological processes in flexible manufacturing systems, such as machining, finishing, precise assembly, automatic inspection, etc., requires control with the force feedback. This, in its turn, requires the knowledge of the dependence between the loading forces/torques and the current in the motor armature.

Let us describe an experiment on the investigation of loading the manipulation robot with electric drives. The experiment was performed at the Institute for Problems in Mechanics of the Russian Academy of Sciences with robot "Universal-5" (see Section 1.1.3). The experimental equipment consisted of the robot, a horizontal table with a steel sheet on it, steel loads, transducers, a computer, and an oscillograph.

The experiment was carried out according to the following plan. The robot moved the loads of different mass along the steel sheet. For each motion, the current in the motor armature winding was measured, and thus the dependence of the current on the load mass was established. Since the only acting force was the dry friction proportional to the mass of the load, we thus obtained the current vs. force. The friction coefficient was constant because all loads were made of the same material (steel). The diagram of the experimental installation is given in Figure 1.2.4.

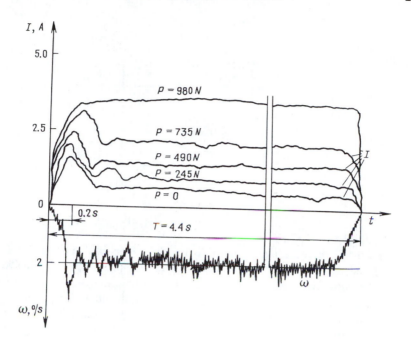

FIGURE 1.2.5.

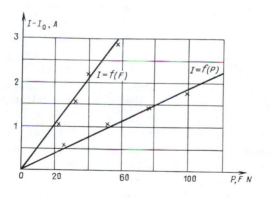

FIGURE 1.2.6.

The acceleration-deceleration curve of the motion of the robot arm had the "triangular" or "trapezoidal" shape (see Figure 1.2.1). The time history of the current I and angular velocity ω for different loads P is shown in Figure 1.2.5. Note that the difference of the current I and the no-load current I_0 is proportional to the load P and the friction force F (see Figure 1.2.6).

In recent years, linear electromagnetic drives have been used in robotic systems, especially for robots performing plane translations.

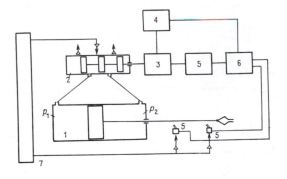

FIGURE 1.2.7.

1.2.3. PNEUMATIC DRIVES

Pneumatic drives are most suitable for the robots with nonservo PTP control systems. Recently, however, new models of pneumatic servo drives and width-pulse modulation systems were developed which are used in more sophisticated control systems such as continuous path, numerical, adaptive, etc. For transport and orientation degrees of freedom, pneumatic drives are mostly used in comparatively simple PTP robots with a load-carrying capacity up to 20 kg. However, grasping motions of grippers are actuated by pneumatic drives in more than 80% of all industrial robots.

Pneumatic drives are used on a large scale in robotic systems due to their simplicity, high speed, and reliability. These drives do not require complex gear trains. They are fire and explosion safe and comparatively cheap. The pneumatic drives can be assembled from typical components, and they can be equipped with pneumatic (fluidic) local control units.

The basic drawback of the pneumatic drives is a significant compressibility of the working medium (air) resulting in the low stiffness of the system. This may cause unwanted oscillations, decrease positioning accuracy, and complicate the creation of the high-quality servo pneumatic systems.

When approaching its extreme positions inside the air cylinder, the piston of the pneumatic drive can strike against the cylinder lids. To avoid shocks and provide smooth braking, one can use hydraulic shock absorbers, feed a back pressure into the cylinder, or attach additional air capacity.[59]

The typical pneumatic robot drive (Figure 1.2.7) comprises a pneumatic cylinder actuator (1), a distribution unit (2), a measuring/control unit (3), a timing unit (4), a sensor and transducer unit (5), a control unit (6), and an air supply unit (7). The drive is controlled by changing the difference of pressures in the "right" and "left" chambers of the air cylinder.

Each unit indicated in Figure 1.2.7 can be of different design. As pneumatic actuators, air cylinders with one or two working chambers, rotary air motors, diaphragm actuators, etc., can be used. The air distribution system may include air valves with different numbers of working positions (usually two or three) and different controls (pneumatic, electromagnetic, mechanical, etc.). The control and timing equipment includes pneumatic resistors and capacitors, fluidic relays, pressure regulators, pressure-operated valves, reference-input elements, pressure indicators, etc. The sensor and transducer unit serves for measuring state parameters of the environment and the robot itself and transforming the primary measurement signals of different physical nature (mechanical, electromagnetic, acoustic, etc.) into pneumatic signals. The pneumatic signals, in their turn, are directly used in the control system for decision making. Note that the robot control system can be constructed on the base of pneumatic modules only, without using electronic devices.

The air supply unit includes air filters and dehumidifiers, lubricant dispersion device, pressure regulators, etc.

In robotics, air cylinders are usually 32 to 80 mm in diameter and the piston stroke makes up to 1 m. The geometrical parameters (piston diameter and stroke, flow areas of pipelines, etc.) and dynamical control modes of the pneumatic drives are calculated starting from the air supply pressure, inertia parameters of the system, and the timing diagram of the robot motion.

The piston of the pneumatic drive moves under the action of the difference of pressures Δp inside the cylinder chambers. This motion is governed by the equation

$$m\ddot{x} = \Delta p \cdot f + \Sigma F_i$$

Here, x is the displacement of the piston; ΣF_i is the sum of all external forces acting on the piston, except the pressure force $\Delta p f$; m is the reduced mass of the driven parts of the manipulator; f is the effective area of the piston. The sum ΣF_i may include friction forces between the piston and the cylinder wall and between the rod and the seal in the guide hole of the cylinder lid, as well as the external forces acting on the manipulator. The pressure force $\Delta p f$ is a control force. The differential pressure Δp is created by the pressure converter included in the control unit.

1.2.4. HYDRAULIC DRIVES

Hydraulic (electrohydraulic) drives are widely used for robots whose load carrying capacities range from 10 kg to several tons. These drives have high speed and hard load characteristics, due to small compressibility of working liquids (mostly different petroleum oils). Basic drawbacks of such drives are possible oil leakage and the necessity to cool working liquids.

Hydraulic motors can be well combined with mechanical gear trains of different design. In the PTP robots, hydraulic cylinders with rectilinearly moving rod and hydraulic piston limited rotary motors are mostly used. For the robots with continuous path control systems, they use the servo drives with linear cylinders or rotary hydraulic motors, and also electrohydraulic stepper drives for translational and rotational motions. The line diagram of the electrohydraulic drives is given in Figure 1.2.8. The drive includes a microcomputer control unit (1), an electromechanical transformer/amplifier

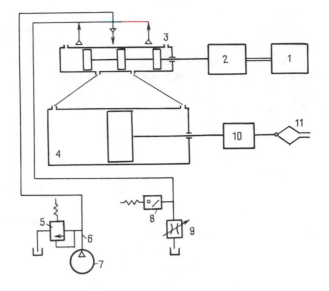

FIGURE 1.2.8.

(2), a distribution unit (3), a hydraulic motor (4), a relief valve (5), a pipeline (6), a hydraulic pump (7), a pressure-operated relay (8), a hydraulic resistor (9), and a position (proximity) sensor (10). The resistor (9) serves for keeping constant the pressure in the corresponding chamber of the cylinder. The number 11 on the diagram indicates the driven manipulator link.

1.3. INFORMATION SYSTEMS OF ROBOTS

1.3.1. PURPOSE AND STRUCTURE

Information systems of robots serve for gathering and processing information about the environment, the manipulation objects, and the robot itself. Information systems are often called sensory systems. The basic purposes of the information system are the reception of primary signals from the robot sensors; transformation of these signals into signals of a different physical nature (electrical, acoustical, optical, hydrodynamical, etc.) if it is required by the control system; preprocessing the primary information and reducing it to a form convenient for the ultimate processing in the control system.

In the general case, the information system (Figure 1.3.1) includes sensors (transducers) (1), preamplifiers (2), analog-to-digital (A/D) and digital-to-analog (D/A) converters (3), and indicators or a monitor (4). The sensors receive primary information about the state of the environment and the robot and convert the information into electric signals. The signals pass through the preamplifier, are transformed by the A/D converter into a digital code and fed to the local and/or global microprocessor control systems. In some cases, local control systems can receive signals from the preamplifier directly,

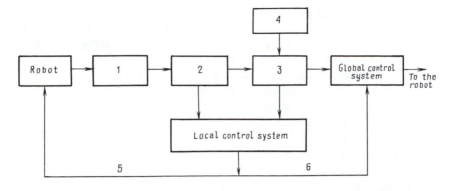

FIGURE 1.3.1.

without A/D transformation. The local and global control systems generate control signals and feed them (through circuits 5 and 6) to the robot actuators. The feedback circuits 5 and 6 shown in Figure 1.3.1 serve also for the communication between the control systems and the information system. If the amount of information is not sufficient for making the decision on how to control the robot, the control system asks for additional information. In this case, the procedure of gathering and processing the measurement data is repeated until sufficient information is obtained.

Concluding this section, we enumerate the basic requirements to the sensors and transducers used for information systems of robots. These are

- high accuracy and time response capabilities (two or three times as high compared with those for drives)
- small mass and dimensions (especially for the sensors and transducers built-in inside grippers)
- the reliability of transmitting the information to the control system and drives
- the combination of transforming and preprocessing the measurement data on each hierarchical level of the information system
- multipurpose applicability of sensor and transducer devices

1.3.2. PARAMETERS TO BE MEASURED; CLASSIFICATION OF INFORMATION SYSTEMS

The variety of robotized technological processes determines the variety of the parameters to be measured. The basic parameters are the linear and angular displacements and velocities of the manipulator links, the distance from the gripper to different objects, the coordinates and velocities of the workpieces inside the working zone, the shape and color of objects, etc. In addition, the robot sensory system can measure such characteristics as clearances, skewnesses, slippage, presence of internal imperfections, coat thickness, material hardness, residual stress in materials, area of different surfaces, orientation of the gripper with respect to the basis axes, etc.

Physical principles of measuring parameters by primary transducers may be different, depending on the physical nature of the observed process, robot design and equipment, accuracy and time response requirements, etc. Modern robots use mechanical, electrical, optical, pneumatic, acoustic, electromagnetic, heat, radiation sensors, etc. In optical sensors, the scanning laser beam is often used. The best result is achieved by combining sensors and transducers of various types in the same robotic system.

The information systems analyze data about the environment and the state of the robot mechanisms. The environment information is necessary for widening the performance capabilities of robots when accomplishing technological tasks, preventing the robot dealing with the technological equipment from breakdowns, and providing a safe work environment for personnel.

Information about the internal state of robots is used to improve their dynamic behavior and for malfunction diagnosis. Using the position and velocity sensors in the control system feedback circuits increases motion smoothness and accuracy. For instance, to achieve nonoscillatory behavior of the robot during acceleration and deceleration periods (see Figure 1.2.1), continuous measuring of positions and velocities of the robot links is, as a rule, required. To increase the reliability and longevity of robots, the automatic inspection of the robot mechanisms state, the reliability prediction, and the emergency shut-down are envisaged in some modern robotic systems. Information (sensory) systems of robots can be classified according to their purposes and the types of primary information, the types of transducers, and the sets of the measured parameters.

By their purposes and types of primary information, sensory systems are classified into

- artificial (computer) vision systems
- detection and ranging sensory systems
- force/torque and tactile sensory systems

The types of transducers can be classified according to their physical principles.

Artificial vision systems allow robots to find different objects and recognize them by their shapes, color, positions and relative orientation, thermal state, surface features, etc. In computer vision systems, optical, radiation, heat, radio wave, and acoustic transducers can be used for receiving and transforming primary information. It seems promising to use holographic and fiber optic devices in computer vision systems.

Since computer vision systems deal with a great amount of information, sophisticated software is needed for processing. At present, many task-oriented software packages have been developed which allow receiving of information through 16 or more channels, carry out a computer analysis (correlation, spectral, etc.) of the image, and represent the information in a generalized form. The output information can be used directly by an operator (in supervisory control) or transmitted to the main processor of the control system.

Note that three-dimensional computer vision systems are beginning to be used in robotics. This seems to be rather promising.

Detection and ranging sensory systems can detect workpieces and obstacles inside the working zone of the robot and measure distances to different objects and velocities. In these systems, the reception and transformation of the measurement information are performed by the transducers based on different physical principles: mechanical, pneumatic, acoustic, electromagnetic, optical, radiation, heat, etc.

Force/torque and tactile sensory systems allow measuring of force/torque, pressure, slippage, mass, temperature, etc. For the reception and transformation of information they mostly use electrical, acoustic, pneumatic, and vibration-frequency methods.

The internal information sensors measure the parameters important for monitoring normal work of the robots and robotic systems and making malfunction diagnosis. These parameters are clearances, linear and angular displacements, velocities, temperature, etc. For coding and transmitting the measurement data such devices as encoders, selsyns, wire potentiometers, inductance transducers, inductosyns, optoelectronic potentiometers, synchro resolvers, etc., are used.

All sensory elements for robots must have high sensitivity and resolution, minimal response time, and be compact and reliable. These requirements provoke the development of new types of transducers, in particular local proximity sensors and artificial vision transducers, including stereoscopic and color measurement devices.

Automatic measurement systems are used in robotic installations involved in such technological processes as assembly, casting, arc and spot welding, forging, electrochemical facing, etc. The correspondence between technological operations and types of transducers is presented in Table 1.3.1. The correspondence between the parameters to be measured and recommended measurement techniques is given in Table 1.3.2. This table also gives typical ranges for the parameters to be measured.

Concluding this section, let us mention sensory systems based on the scanning laser beam. Though such systems have not yet found a wide application in robotics, they seem to be promising for special robotic installations for measurement and diagnostics. Laser scanning is used now in some prototypes of industrial robots.

In the next section, we consider in more detail pneumatic sensory systems, which have a number of advantages that make them preferable in some applications.

1.3.3. PNEUMATIC MEASURING SYSTEMS

Pneumatic measuring systems are simple, reliable, durable, and comparatively cheap. They are appropriate for use in robots and robotic systems

TABLE 1.3.1.
Technological Processes and Relevant Transducers

Technological process	Operation	Transducer type
Assembly	Object recognition	TV camera, fluidic and ultrasonic transducers, photocells
	Inserting a peg into a hole; recognition of a hole	Electromechanical, elastic, pneumatic
	Workpiece gripping	Spring (mechanical and pneumatic), piezoelectric
	Orientation of the gripper and manipulation object	Mechanical, electromagnetic
Casting	Inspection of the moulding sand composition	Optical, ultrasonic
	Checking the temperature of casting metal	Thermoresistors
	Measurement of the casting dimensions	Photocells, fluidic, and ultrasonic transducers
Welding	Measurement of the distance to the surface; detection of the workpiece and the weld	Fluidic, optical, ultrasonic, electromagnetic
	Detection of weld imperfections	Eddy current, ultrasonic
	Detection of edges	Eddy current, fluidic
Forging	Shape determination	Ultrasonic, fluidic, photoelectric, fiber optical
	Loading/unloading workpieces in presses	On/off switches, fluidic
	Measurement of the mass	Mechanical
	Determination of the grasping force of the gripper	Pneumatic cushions
	Orientation of the workpiece	Fluidic
	Position determination	TV camera

intended for stamping, welding, casting, machining, automatic monitoring, etc. Note that recently, some firms began to use pneumatic sensors in industrial robots with simple control systems, in particular, PTP ones. For example, the firm Auto-Place (U.S.) produces industrial robots with pneumatic drives, PTP control systems, and pneumatic sensors for detecting the presence of the workpiece in the gripper. In fact, these sensors measure the distance between the gripper jaws. If the distance is less than a definite value, this indicates the absence of the workpiece in the gripper.

It seems to be promising to use fluidic devices in sensory systems of robots. The response time of these devices is about 10^{-3} s, their power consumption makes up to several watts, and the output power is about 10^{-1} W. These parameters are sufficient for measurement operations. Pneumatic transducers can be successfully used in proximity, force/torque, and tactile

TABLE 1.3.2.
Measurement Techniques for Different Parameters

Parameter to be measured	Measurement range	Measurement error	Measurement technique
Angular displacement	0–180°, 0–360° 0–270°	±5°	Optical, electromechanical
Linear displacement	0.1–0.01 m; 0.1–1.3 m	0.05 mm 0.1 mm	Electromagnetic, potentiometric
Distance to the object	0.02–0.3 m	0.1 mm	Microwave, fluidic, electro-magnetic, optical
Shapes of objects	≤0.3 m		Photoelectric, ultrasonic
Monitoring the motion of an object			Optical, stereoscopic
Investigation of internal structure (inclusions)	≤0.01 m		Ultrasonic, eddy current, X-rays
Detection of object	From 0.01–0.4 m through 2 m	0.1 mm 3 mm	Microwave, electromagnetic, fluidic, optical, ultrasonic
Tracking the contour			Optical, fluidic, eddy current
Roughness	0.01–0.001 m		Optical, fluidic, inductive
Grasping force	Through 200 N	0.1 N	Strain measurement, electro-magnetic
Slippage	0.01–0.05 m	0.01 mm	Electromagnetic
Pressure	Through 10^4 Pa	20 Pa	Strain measurement

sensory systems. These sensors measure such parameters as linear and angular displacements, distances, velocities, accelerations, forces/torques, pressure, temperature, etc. (see Table 1.3.3). The sensors can be located in grippers, joints, and actuators as well as in technological equipment, for example, in mold matrices for forming.

Pneumatic transducers are appropriate for monitoring different parameters of technological processes. Most of the monitoring systems are based on hydrodynamical and gas dynamical principles (see Table 1.3.3). They use such techniques as jet chopping, turbulization of the laminar jet, closing partly the discharging jet, working against a definite force or pressure, changing the fluid oscillator frequency depending on the temperature, and pulse processing in frequency jet choppers. Usually, the measurements are carried out by the difference compensation method.

Let us give the performance characteristics achievable for pneumatic sensory systems, depending on the parameter or property to be measured.

Linear displacement in the near zone (near ranging) — The measurement range is 0 to 4 mm; the minimal and maximal errors are 0.5 and 3%, respectively; the operation time is 0.01 s; the sensitivity is 200 Pa/μm.

Detection of the object — The measurement range is 0 to 20 mm (the pneumatic-acoustic method allows extending the range to 0.5 m); error is 2 to 8%; the operation time is 0.02 s; the sensitivity is 200 Pa/μm.

TABLE 1.3.3.
Principles and Techniques of Measurement with
Pneumatic Transducers

Parameter to be measured	Physical principles and measurement techniques	Diagram
Presence of workpieces and other objects	Jet chopping, turbulization of the laminar jet	
Small displacements and distances	Closing partly the discharging turbulent jet	
Linear dimensions of objects	Oscillation generation, pulse counting: $l = f(\omega)$	
Pressure, force	Work against a certain force: $p_1 = k \cdot p_2$	
Temperature	Temperature dependence of the fluidic generator frequency	
Number of revolutions, angular velocity	Jet chopping, counting interruption pulses	
Thickness of sheets	Difference pressure measurement: $h = f(\Delta p)$	

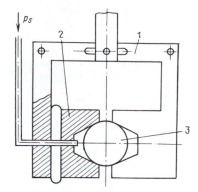

FIGURE 1.3.2.

Force — The measurement range is 0.1 to 10 N; the operation time is 0.2 s.

Temperature — The measurement range is 10 to 800°C; the error is 5 to 10%; the operation time is 20 to 30 s; the resolution is 10°.

Pressure — The measurement range is 0 to 10 MPa; the error is 2 to 5%; the operation time is 0.2 to 1 s; the resolution is 0.1 MPa.

The advantages of pneumatic sensory systems for industrial robots are their small dimensions and mass, reliability, and simplicity of production and operation. They can work in intensive electromagnetic fields (for example, in welding robots) without distortions. They can operate in high-radiation zones, under intensive shocks and vibrations, in inflammable and explosive media, and in other extreme conditions.

As drawbacks of pneumatic sensory systems, one should mention a comparatively short range and low speed and large dimensions of power supply equipment compared with those for other sensory devices.

For some types of industrial robots, fluidic sensors which measure the workpiece size are built-in inside the gripper. The fluidic elements serve for determining the position of the workpiece in the gripper and for control of the grasping mechanism. The gripper with a built-in sensor is shown in Figure 1.3.2 where 1 is the gripper itself, 2 is the fluidic sensor (measurement transducer), and 3 is the workpiece. The sketch of a fluidic measurement transducer is given in Figure 1.3.3. Here, p_s is a power supply pressure, p_0 is the atmospheric pressure, p_1 and p_2 are pressures at output ports, and p_i is the pressure inside the jet interaction chamber.

Summing up, let us enumerate the preferable fields for application of pneumatic sensory devices in industrial robots and robotic equipment. Pneumatic sensory systems should be used:

1.   In robots with pneumatic drives, for detecting objects, measuring distances and small displacements

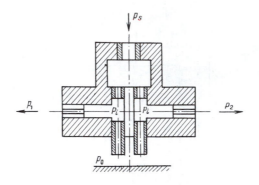

FIGURE 1.3.3.

2.   In robots operating in extremal conditions (fire- and explosion-hazard-
     ous media, high temperature, radiation, intensive electromagnetic fields,
     etc.), for obtaining maximum trustworthy information about the ex-
     amined parameters
3.   In "sensory grippers" for industrial robots, irrespective of types of
     actuators and operation conditions

In Chapter 8 we give more detailed information about pneumatic measurement systems and their application in robotic systems for measurements
and inspection (see also References 96–98, 101, 103, 104, 110, and 111).

1.4. CONTROL SYSTEMS OF ROBOTS

1.4.1. MAIN FUNCTIONS AND BASIC TYPES OF CONTROL SYSTEMS

The control system plans the robot motion according to a prescribed
operation task and the information received from the sensory system generates
control signals and engages the actuators to implement the planned motion.
Depending on the way of specifying (planning) robot motions, control systems
can be classified into PTP, programmed sequence, continuous path, and combined control systems.

For PTP control systems, only the sequence of positions of the gripper
and (sometimes) the links are specified which the robot has to pass when
implementing the motion. In the simplest case, only the final position is
indicated. For programmed sequence control systems, the time table of passing
the prescribed positions is, in addition, specified. Very often, PTP and programmed sequence control systems are used in robots aimed at implementing
motions between two positions repeatedly. The control systems designed
specially for such robots are sometimes called cycle control systems.

For continuous path control systems, the motion planning consists in
specifying the continuous trajectory which the robot can track when implementing the operation. Note that for some operations it is sufficient to indicate

only the geometrical curve, while for others the pace of moving along this curve is also important and must be specified. Continuous path control systems are more sophisticated than PTP and programmed sequences ones and, as a rule, use a computer or a microprocessor for on-line calculating control parameters.

Combined control systems allow use of both PTP/programmed sequence and continuous path modes.

Control systems are also classified according to the code (analog or numerical) used for generating control signals, program medium (magnetic tape or disk, punched tape, master cams, stops, etc.), programming mode (on- or off-line, automatic, interactive, teaching, etc.), the presence/absence of feedback circuits and the content of feedback information, and hardware components used for information processing in the control system (electronic, mechanical, pneumatic, etc.).

In modern control systems of robots, electronic hardware components are generally used. However, in some applications, other devices can be preferable, depending on the types of actuators and sensors used in the robot. For example, if the robot is equipped with pneumatic drives and a pneumatic sensory system, pneumatic (fluidic) components can also be used in its control system.

The open-loop control systems are the simplest ones. They need no current information about the environment and robot motion and work following the "rigid" program. Such control systems are applicable only in cases where environmental and technological parameters are kept constant. The majority of robot control systems include feedback circuits.

To achieve high-performance capabilities (in productivity, accuracy, repeatability, reliability, etc.) for robots operating in a complex environment and carrying out complex technological operations, a great amount of information should be processed. The control system of many modern robots, especially of those used in computer integrated manufacturing (CIM) and flexible manufacturing systems (FMS) have a hierarchical structure which allows processing of information in the most rational way. Such control systems can work in an automatic mode entirely or with the interference of a man-operator in some situations.

Control systems must correspond to the technological purposes of robots and their operation conditions. While the control system with a sophisticated structure and great amount of feedback is necessary for tasks in a complex environment under changing conditions, in other situations simpler control systems are preferable. In particular, robots with PTP/cycle and programmed sequence control systems find a wide application.

The cycle control system (Figure 1.4.1) comprises a control console (1), a program storage unit (2), a command generation unit (3), a control unit (4), final amplifiers (5), and an interface unit (6) connecting the control system with sensors. The only sensors necessary for operation of the cycle control system are proximity sensors which inform the robot about the approach of

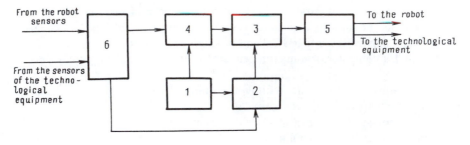

FIGURE 1.4.1.

the gripper to terminal positions. The sensors can be placed on the robot and/or on the technological equipment. For the simplest nonservo cycle control system, instead of the proximity sensors, on/off switches are used which disengage actuators when the gripper approaches the terminal position. Examples of robots equipped with nonservo cycle control systems are Tsiklon and Ritm (see Section 1.1.3). In References 80 and 111, the pneumatic cycle control system is described, intended for robots serving presses and forging equipment. This system uses fluidic modules whose work is based on turbulization of jets discharging from a capillary. Nonservo PTP cycle control systems are used in robots whose operations do not require high accuracy, when the environmental conditions and parameters of the robot practically undergo no changes. However, due to gradual wear of the robot parts, gear trains and other mechanisms acquire backlashes, friction in bearings changes, etc. This deteriorates the dynamic behavior and accuracy of the robot. To avoid decrease of positioning accuracy, the position feedback circuit is introduced in the control system, and servo drives are used. Examples of robots with PTP servo control systems are TUR-10 and Universal (Russia).

Continuous path control systems[75,112,172] can ensure a sophisticated motion of the robot end-effector along the planned trajectories. Examples of robots with continuous path control systems are Puma® (U.S.), Skilam (Japan), RPM-25, TUR-10K (Russia), etc.

Continuous path control requires processing a great amount of information and can be implemented by using computers with large memory capacity. The block diagram of one version of the continuous path control system is shown in Figure 1.4.2. Information about the environment and robot state comes from position and velocity sensors, technological equipment sensors, and emergency on/off switches. The control system envisages the following operation modes: PLAYBACK (automatic tracking a specified trajectory), TEACHING (programming from the terminal), MANUAL CONTROL (control from the operator console), and CHECK (inspection of the equipment and malfunction diagnostics). The desired trajectory of the robot can be specified by entering the trajectory parameters from the program terminal (when operating in TEACHING mode) or by "manual teaching". In the latter case the gripper and/or the links of the robot are manually led along the

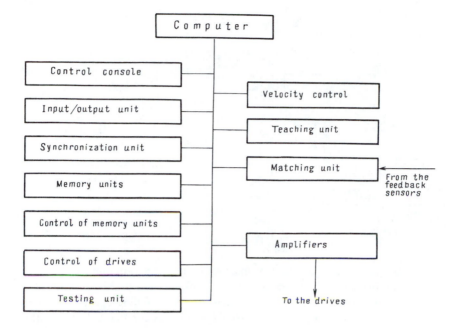

FIGURE 1.4.2.

desired trajectory, the information about the trajectory shape being automatically entered into the program unit. The positioning accuracy of robots with servo PTP and continuous path control systems makes up 0.5 to 1.0 mm. For the most precise robots (for example, Skilam, see Figure 1.1.12) the accuracy achieves 0.05 mm.

1.4.2. EXTENSION OF CONTROL CAPABILITIES — ADAPTIVE CONTROL SYSTEMS

To improve performance capabilities of robots, they must be equipped with developed information and a feedback control system. If the working conditions are uncertain and change unpredictably, the artificial intelligence and adaptive control systems should be used which can automatically alter the control program so as to match the environmental changes. For these control systems, special computer devices such as sensory processors or image processors are developed which can be integrated with corresponding transducers.

Let us consider the control system combined with the information system using inertial (gyroscopic) sensors built-in inside the gripper. This control system was designed for industrial robots transporting massive loads and/or performing precise technological operations such as assembly, laser or arc welding and cutting, inspection of technological equipment, etc. The block diagram of the control system is presented in Figure 1.4.3. It includes a manipulator (1); inertial sensory elements (2 to 4) placed in mutually or-

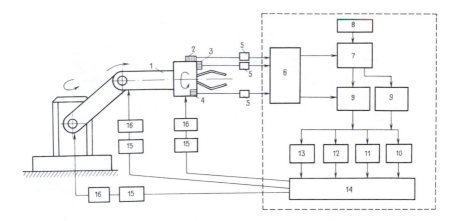

FIGURE 1.4.3.

thogonal planes; transducers (5); microprocessor (6); program medium (7); programming terminal (8); commutators (9); units (10 to 13) for control of displacement, velocity, acceleration, and torque; matching unit (14); driving motors (15); and reduction gears (16). The sensors for internal information are not shown in Figure 1.4.3. Signals from the inertial sensors (2 to 4) are transmitted through the transducers (5) into the computer unit (6). The computer processes the input data and calculates the control signals for correction of the programmed motion. In the control system in question, the position and orientation of the robot gripper are measured directly by the built-in sensors inside the gripper itself. This allows determination of the current position of the gripper with high accuracy. In the systems of traditional design, the position sensors are usually placed only at the joints, and the accuracy of determining the position of the gripper is reduced due to joint backlashes and elastic compliance of the joints and links of the robot.

The described control system can accomplish various tasks: quenching oscillations in actuators, keeping up the end-effector position and orientation, precisely tracking a prescribed trajectory, etc. Note that apart from the gyroscopic sensors, other measuring devices can be also used, e.g., laser measurement devices.

Control systems of robots operating in a complex environment require sophisticated multilevel software that includes data processing, trajectory planning, and decision making algorithms which ensure adaptive, intelligent behavior of the robot. Adaptive and intelligent control systems, as a rule, have a hierarchical structure, including task forming, motion planning, and decision-making and control signal-generating levels. The decision-making and motion-planning levels interact with the sensory system and data processing algorithms delivering the current information about the environment and robot state. In addition, the interference of a man-operator into the control process can be envisaged.

Chapter 2

BASIC KINEMATICS AND DYNAMICS OF MANIPULATION ROBOTS

2.1. MODELS AND PROBLEMS OF ROBOT DYNAMICS

2.1.1. INTRODUCTION

Modern manipulation robots are complex engineering systems used for automatizing different technological operations: transport, assembly, welding, cutting, measurement, etc. A manipulation robot consists of a control system, an information system, and a mechanical system (a manipulator itself). The control system serves for programming, storing, and implementing a control code. The information system gathers current data about the state of different units of a robot (for instance, the values of relative displacements and velocities of manipulator arm links) and the environment, preprocesses these data and transmits them into the control system. Using these data, the control system generates commands which determine voltages, currents, or pressures in corresponding actuators. The mechanical system (the manipulator itself) performs prescribed motions, thus fulfilling the technological task of the robot.

Usually, the arm is a system of links connected by kinematic pairs. Manipulation robots use mostly kinematic pairs of the fifth class (both revolute and rectilinear), which provide a single degree of freedom for relative motion of each couple of adjacent links. The class of a kinematic pair is defined as the number 6-k, where k is the number of degrees of freedom for relative displacement of the links connected by this pair.

A revolute kinematic pair of the fifth class (a revolute joint) admits relative rotation of links (joined by the pair) about their common axis. A rectilinear kinematic pair of the fifth class, or prismatic joint, allows corresponding links to move, relative to each other, along a specified straight line. Apart from the above mentioned kinematic pairs, kinematic pairs of higher classes are used in manipulation robots as well, spherical joints (third-class kinematic pairs providing relative rotations of links with respect to three axes) and two-degree-of-freedom joints (fourth-class kinematic pairs admitting relative rotations of links with respect to two axes) being the most widely used. However, when describing manipulator kinematics, we can treat these joints as a combination of three or two revolute joints, respectively, and thus limit ourselves to kinematic pairs of the fifth class.

More detailed consideration of the manipulation robot design can be found in References 15, 24, 87, 145, 178, 186, 187, 193, 258.

In what follows we often use the terms "a mechanical system of a robot", "a robot", "a manipulation robot", and "a manipulator" as synonyms.

2.1.2. MECHANICAL MODELS

Analysis of manipulation robot dynamics and control requires adequate models describing mechanical properties of real robots with sufficient accuracy. In each specific case, the choice of the model is determined by the kinematics of the manipulator, its mechanical parameters (inertial, elastic, dissipative, etc.), types and characteristics of actuators, as well as by the necessary accuracy of the model.

In the simplest models, all components of a manipulator (parts of its carrying structures, arm links, gears and shafts, rotors of electric motors, etc.) are considered to be perfectly rigid bodies. Kinematic pairs and gear trains are treated as ideal. If transmission mechanisms include flexible elements, such as chains, belts, or flexible shafts, their inertia is neglected in these models. Thus, we assume that the gears provide ideal geometric constraints between moving rigid parts of the manipulator. Such models reflect basic properties of many real manipulators with acceptable accuracy and are often used in mechanics of robots.

However, in some cases it is necessary to consider more sophisticated models. For example, significant oscillations of a manipulator may arise due to elastic compliance of its parts. These oscillations can cause considerable deterioration of the manipulator performance and reduce the positioning accuracy. To analyze such oscillations and develop efficient methods for their damping, it is necessary to take into account elastic properties of the robot links, joints, and gears. Sometimes it is also necessary to take into account friction and backlashes in kinematic pairs and gears.

Dynamics of actuators (electric, pneumatic, or hydraulic) may also require more detailed consideration. In this case, the mechanical model of the robot must be supplemented with equations describing electromagnetic, aero-, or hydrodynamical processes.

Thus, mathematical models of robots can be presented as sets of differential equations. Note that for the same robot different models can be used, depending on the specific purpose of modeling and the desired accuracy of description.

Let us consider possible mechanical models for a manipulator of the simplest design, shown in Figure 2.1.1. The manipulator consists of a base (1), a direct current (DC) electric motor with independent excitation (2), a reduction gear train (3), a single-link arm (4), and a gripper (5), rigidly attached to the arm. The motor body is fixed on the base. The drive gear (6) is rigidly connected with the shaft (7) of the electric motor rotor (armature). The driven gear (8) and the arm (4) are rigidly fastened to the shaft (9). This shaft is pointed vertically and rotates in bearings placed on the base.

The simplest mechanical model of such a manipulator is a single-degree-of-freedom system, viz., a rigid body (the group comprising the arm (4), the shaft (9), and the driven gear (8)) rotating about a fixed axis under the action of a control torque M generated by an actuator and transmitted through the

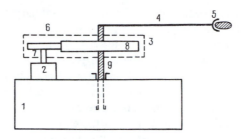

FIGURE 2.1.1.

gear (6). The motion of the system is governed by the second-order differential equation

$$I\ddot{\varphi} = M \qquad (2.1.1)$$

where φ is the rotation angle of the arm, I is the total moment of inertia of the arm (4), the driven gear (8), and the shaft (9), with respect to the rotation axis.

Having taken into account the elastic (torsional) compliance of the shaft (9), we arrive at the following equations of motion:

$$I_1 (\ddot{\varphi} + \ddot{\psi}) = M, \ c\psi = -M. \qquad (2.1.2)$$

In Equations 2.1.2 the mass of the shaft (9) as well as the mass of the gear (8) is neglected. Here, φ is the rotation angle of the gear (8) with respect to the base, ψ is the angle of relative rotation of the arm with respect to the gear (8) caused by the elastic compliance of the shaft (9), I_1 is the moment of inertia of the arm with respect to the rotation axis, and c is the torsional stiffness coefficient of the shaft. The first equation in Equations 2.1.2 describes the change of the angular momentum of the arm with respect to the rotation axis while the second one is the equilibrium condition for the gear (8) under the control and elastic torques. Note that the second equation in Equations 2.1.2 is a finite relationship because the mass of the gear (8) is assumed to be zero. Equations 2.1.2 formally turn into Equation 2.1.1 as $c \to \infty$, provided the torque M is restricted and $I_1 = I$.

Let us take into account the moment of inertia of the gear (8), the inertia of the shaft (9) being still neglected. Then we obtain the following modification of the manipulator mechanical model:

$$I_1 (\ddot{\varphi} + \ddot{\psi}) + I_2\ddot{\varphi} = M, \ I_2\ddot{\varphi} = M + c\psi \qquad (2.1.3)$$

Here, I_2 is the moment of inertia of the gear (8) with respect to its axis. Equations 2.1.3 turn into Equation 2.1.2 when $I_2 = 0$.

The models considered above are described by ordinary differential Equations 2.1.1 to 2.1.3. Let us give an example of the model including partial differential equations. Let the manipulator arm (4) be a thin elastic beam of a constant circular cross section while the driven gear (8) and the shaft (9) be perfectly rigid. When undeformed, the arm is rectilinear and orthogonal to the rotation axis. We introduce the following denotations: φ is the angle of rotation of the shaft (9); x is the coordinate measured along the axis of the arm; $v_1(x, t)$ and $v_2(x, t)$ are projections of the elastic displacement vector of the corresponding point of the arm onto the horizontal and vertical planes, respectively, at the time instant t; ρ is the linear density of the arm; J is the moment of inertia of the arm cross section with respect to its diameter; E is Young's modulus of the arm material; l is the length of the arm; g is the gravity acceleration. Usually, links of manipulation robots are rather stiff. Therefore, elastic displacements of the links are small, and their natural frequences are high compared with their angular velocities. Under these conditions, linear theory of elastic beams can be used. Thus, the motion of the manipulator arm is described by linear partial differential equations with boundary conditions

$$\rho\frac{\partial^2 v_1}{\partial t^2} + EJ\frac{\partial^4 v_1}{\partial x^4} = -\rho x\ddot{\varphi} \qquad \rho\frac{\partial^2 v_2}{\partial t^2} + EJ\frac{\partial^4 v_2}{\partial x^4} = -\rho g \qquad (2.1.4)$$

$$I\ddot{\varphi} + \int_0^l \rho\frac{\partial^2 v_1}{\partial t^2}x \, dx = M \qquad (2.1.5)$$

$$v_i(0, t) = \frac{\partial v_i(0, t)}{\partial x} = 0, \quad \frac{\partial^2 v_i(l, t)}{\partial x^2} = \frac{\partial^3 v_i(l, t)}{\partial x^3} = 0, i = 1, 2 \quad (2.1.6)$$

The partial differential Equations 2.1.4 describe oscillations of the arm under the action of elastic, gravity, and inertia forces. The integrodifferential Equation 2.1.5 governs change of the angular momentum of the system. The boundary conditions (Equations 2.1.6) mean that one end of the arm is fixed to the shaft while its other end is free.

Let us substitute the expression for $\rho\partial^2 v_1/\partial t^2$ from the first Equation 2.1.4 into 2.1.5 and calculate the integral, taking into account boundary conditions (Equations 2.1.6). Thus we obtain

$$\left(I - \frac{\rho l^3}{3}\right)\ddot{\varphi} = M + EJ\frac{\partial^2 v_1(0, t)}{\partial x^2} \qquad (2.1.7)$$

Equation 2.1.7 describes the motion of the rigid part of the system (the driven gear (8) and the shaft (9)) under the torque M generated by the actuator, and the torque $EJ\partial^2 v_1(0, t)/\partial x^2$ of elastic interaction between the shaft and the arm. The coefficient $(I - \rho l^3/3)$ is the total moment of inertia of the gear (8) and the shaft (9). Sets of Equations 2.1.4, 2.1.5, 2.1.6, and 2.1.4, 2.1.7, 2.1.6 are equivalent.

Consider now a somewhat more sophisticated model of the robot (see Figure 2.1.1) including dynamics of its DC actuator.

The angles χ and φ of rotation of gears (6) and (8), respectively, are related by

$$\chi = -n\varphi \tag{2.1.8}$$

where n is the gear ratio. The minus sign in Equation 2.1.8 shows that the gears (6) and (8) rotate in opposite directions. For a DC motor with independent excitation the following relationships hold:

$$L\, dj/dt + Rj + k_1\dot{\chi} = u, \quad \mu = k_2 j \tag{2.1.9}$$

where L and R are the inductance and resistance of the armature winding, j is the armature circuit current, u is the control voltage, μ is the electromagnetic torque, and k_1 and k_2 are constant coefficients, depending on the voltage in the input of the excitation circuit. Having eliminated j from Equation 2.1.9 and using Equation 2.1.8 we obtain

$$L\dot{\mu} + R\mu - k_1 k_2 n\dot{\varphi} = k_2 u \tag{2.1.10}$$

The equation of motion for the rotor, together with the drive gear (6), has the form:

$$I_0\ddot{\chi} = \mu + M_1 \tag{2.1.11}$$

Here, I_0 is the total moment of inertia of the rotor and the gear (6), M_1 is the torque with which the gear (8) acts on the gear (6). The torque M_1 and the torque M with which the gear (6) acts on the gear (8), are connected by the relation

$$nM_1 = M \tag{2.1.12}$$

Equation 2.1.12 follows from the fact that the torques M and M_1 are proportional to the radii of the respective gears (8) and (6), the ratio of these radii being equal to n. From Equations 2.1.8, 2.1.11, and 2.1.12 we obtain

$$M = -n\mu - I_0 n^2 \ddot{\varphi} \tag{2.1.13}$$

Thus, the DC actuator can be described by Equations 2.1.10 and 2.1.13. If we add these equations to the models in Equation 2.1.1 through 2.1.7 considered above, we arrive at the models taking account of actuator dynamics.

2.1.3. BASIC PROBLEMS OF ROBOT DYNAMICS

Equations of a robot as a dynamic controlled system usually can be presented in the following general form:

$$\dot{x} = f(x, u, t), \ t_0 \le t \le T, \ x = (x_1, \ldots, x_n),$$
$$u = (u_1, \ldots, u_m), \ f = (f_1, \ldots, f_n) \tag{2.1.14}$$

Here, x is the n-vector of phase (state) coordinates; u is the m-vector of control variables; f is the n-vector function, depending on the design and dynamic properties of a manipulator; and t_0 and T are the initial and terminal time instants. Initial conditions for the system in Equation 2.1.14 are given by

$$x(t_0) = a \tag{2.1.15}$$

The state vector x in Equation 2.1.14 comprises generalized coordinates and velocities of a manipulator (angles, linear displacements, angular and linear velocities) and some other variables related to dynamics of actuators (e.g., electric currents). The control vector u may include forces and torques created by actuators, control voltages, etc. For instance, in the model described by Equation 2.1.1 we have $x = (\varphi, \dot{\varphi})$, $u = M$, while in the models described by Equations 2.1.2 and 2.1.3 $x = (\varphi, \psi, \dot{\varphi}, \dot{\psi})$, $u = M$. The model described by Equations 2.1.4 through 2.1.6 includes partial differential equations and is not presented in the form of Equation 2.1.14. Here, $u = M$ while the state is described by functions $v_1(x, t)$, $\partial v_1(x, t)/\partial t$, $v_2(x, t)$, $\partial v_2(x, t)/\partial t$, $\varphi(t)$, $\dot{\varphi}(t)$. If the above mentioned models are supplemented with equations of the electric actuator (Equations 2.1.10 and 2.1.13), the variable μ is added to the state vector x, and the voltage u plays the role of control variable.

Let us state two basic problems of robot dynamics.

Problem 2.1. Determine the motion of the manipulator under initial conditions (Equation 2.1.15) and the control law $u = u(x, t)$.

Problem 2.1 is a generalization of the direct problem of classical mechanics: find the motion of the system under given forces and initial conditions.

Note that the function $u = u(x, t)$ specifies a feedback control law. If, in particular, u is independent of x, we have an open-loop (programmed) control law, $u = u(t)$.

In fact, solving Problem 2.1, i.e., integrating Equations 2.1.14 with initial conditions (Equation 2.1.15), is nothing but mathematical simulation of the robot motion. Usually, Equations 2.1.14 are nonlinear and require numerical integration. In some cases application of approximate analytical (asymptotic) methods turn out to be efficient.

Problem 2.2. Let some conditions (requirements) C be imposed on the robot motion, and an admissible set U of control functions ($u \in U$) be given. Find admissible controls ($u \in U$) satisfying conditions C.

Problem 2.2 is a generalization of the inverse problem of classical dynamics: given a motion of a mechanical system, find the acting forces.

Problem 2.2 covers a wide class of robot mechanics problems, which differ by the imposed conditions C. For instance, we may impose various constraints on the robot motion (geometrical, operational, technological, etc.), stability requirements, optimality conditions with respect to time, energy, accuracy, or some other performance criterion, etc. Typical examples of Problem 2.2 are those of stabilization of a planned trajectory and optimal control.

The problem of stabilization can be stated as follows. Given the desired motion $x = x_0(t)$ for $t \in [t_0, T]$, find such linear feedback control $u(x, t) = K(t)x$ that for any motion x(t) satisfying the condition $\|x(t_0) - x_0(t_0)\| < \delta < \epsilon$, where $\delta > 0$, $\epsilon > 0$ are given numbers, the inequality $\|x(t) - x_0(t)\| < \epsilon$ holds for all $t \in [t_0, T]$. Here, $\|z\|$ denotes the norm of the vector z, and the $m \times n$ time-depending matrix $K(t)$ is to be determined.

A rather general optimal control problem can be stated as follows. Find the admissible control $u^*(x, t) \in U$ and the corresponding motion $x = x^*(t)$ of the system governed by Equation 2.1.14 for $t \in [t_0, T]$ such that:

1. The initial state vector belongs to the given set S_0

$$x^* (t_0) \in S_0 \qquad (2.1.16)$$

2. The motion satisfies constraints imposed on the state and control:

$$\{x^* (t), u^* (x(t), t)\} \in G(t), \quad t \in [t_0, T] \qquad (2.1.17)$$

(here, G is the given set in $(n + m)$-dimensional space which can depend on time).

3. The terminal state belongs to the given set S_T:

$$x^* (T) \in S_T \qquad (2.1.18)$$

4. The functional J, depending on the state and control, is minimal (or maximal).

The set S_0 in Equation 2.1.16 may, for instance, describe possible initial configurations of a manipulator while S_T in Equation 2.1.18 may denote a

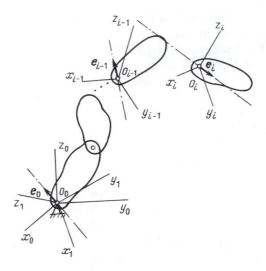

FIGURE 2.2.1.

neighborhood of the desired terminal state. The set G may express constraints imposed on control voltages, electric currents, and admissible configurations of the robot. Constraints in Equation 2.1.17 may take account of obstacles as well as velocity restrictions. The functional J reflects the performance quality of a manipulator, for instance, the time T of accomplishing a technological task, total energy consumption, heat losses in the electric motor, etc. The specific form of the functional J is determined by a technological purpose of a robot and its working conditions.

Various kinds of Problems 2.1 and 2.2 are considered in subsequent chapters of this book.

2.2. KINEMATIC RELATIONSHIPS

The kinematic structure of a manipulation robot comprises a base and links of an arm. A manipulator arm, together with a base, forms an open kinematic chain, i.e., the set of bodies connected consecutively by kinematic paris (Figure 2.2.1). We assume all kinematic pairs to be either revolute joints or prismatic joints. Spherical (three-degree-of-freedom) and two-degree-of-freedom joints are presented as a combination of three or two revolute joints, respectively. Rotation axes of revolute joints and guide lines of prismatic joints will be called axes of corresponding kinematic pairs.

Denote by n the number of manipulator arm links. Let us number the links consecutively from 1 to n, starting from the link attached to the base and assigning the index 0 to the base itself. The unit vector of the sth kinematic pair (connecting $(s - 1)$th and sth links) axis will be denoted by $\mathbf{e}_s$, $s = 1, \ldots, n$.

Let us connect a right-hand Cartesian coordinate frame $O_s x_s y_s z_s$ ($s = 0$, $1, \ldots, n$) with the sth link of the manipulator, placing the origin O_s on the axis of the sth kinematic pair. If the sth link of the manipulator is a perfectly rigid body, the corresponding coordinate frame is fixed with respect to the link. If the link is deformable, we connect the coordinate frame $O_s x_s y_s z_s$ with a fictitious rigid body corresponding to the undeformed state of the link. The vector $\mathbf{e}_s$ is fixed in the coordinate frame $O_s x_s y_s z_s$. As a rule, the frame $O_s x_s y_s z_s$ is chosen so that one of the coordinate axes is parallel to $\mathbf{e}_s$. If the sth kinematic pair couples two perfectly rigid links, the vector $\mathbf{e}_s$ is fixed in both frames $O_{s-1} x_{s-1} y_{s-1} z_{s-1}$ and $O_s x_s y_s z_s$.

To describe a robot motion we introduce the set of generalized coordinates q determining the manipulator configuration. Cartesian coordinates of each point of the manipulator in any reference frame $O_s x_s y_s z_s$ ($s = 0, 1, \ldots, n$) can be expressed through the coordinates q.

If all links of a robot are rigid, the generalized coordinates q usually include angles of relative rotations for revolute joints, and linear relative displacements for rectilinear kinematic pairs. If a manipulator contains deformable (elastic) parts, generalized coordinates also include variables and/or functions describing deformation of joints and/or links.

The basic problem of robot kinematics can be formulated as follows.

Problem 2.3. Find coordinates, velocity, and acceleration of an arbitrary point A of the robot in the prescribed coordinate frame $O_s x_s y_s z_s$ expressed in terms of generalized coordinates q of the robot, velocities $\dot{q}$, and accelerations $\ddot{q}$. Coordinates of the point A in some reference frame $O_k x_k y_k z_k$ are assumed to be known.

The solution of Problem 2.3 will make use of transformation formulas for coordinates, velocities, and accelerations. Below we present some of these transformations in the form convenient for applications.

Denote by $\{\mathbf{a}\}_s$ the column vector comprising coordinates a_{x_s}, a_{y_s}, a_{z_s} of an arbitrary vector $\mathbf{a}$ in the reference frame $O_s x_s y_s z_s$:

$$\{\mathbf{a}\}_s = \|a_{x_s}, a_{y_s}, a_{z_s}\|^T$$

Here and in what follows, the superscript "T" denotes the transposition of a vector or a matrix.

Denote by $\dot{\mathbf{a}}^{(s)}$ the time derivative of a vector $\mathbf{a}$ with respect to the reference frame $O_s x_s y_s z_s$. By definition, the components of the vector $\dot{\mathbf{a}}^{(s)}$ in the reference frame $O_s x_s y_s z_s$ are time derivatives of components of the vector $\mathbf{a}$ in this reference frame:

$$\{\dot{\mathbf{a}}^{(s)}\}_s = d\{\mathbf{a}\}_s/dt = \|\dot{a}_{x_s}, \dot{a}_{y_s}, \dot{a}_{z_s}\|^T \qquad (2.2.1)$$

Consider two coordinate frames, $O_i x_i y_i z_i$ and $O_j x_j y_j z_j$ (Figure 2.2.1). Let $\mathbf{R} = \overline{O_i O_j}$ be the radius-vector of the point O_j with respect to the point O_i; let

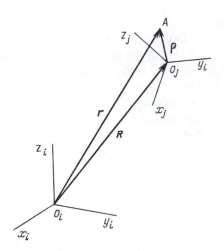

FIGURE 2.2.2.

T_{ij} be the matrix of direction cosines for the axes of the frame $O_i x_i y_i z_i$ with respect to the frame $O_j x_j y_j z_j$:

$$T_{ij} = \begin{Vmatrix} \cos(x_i, x_j) & \cos(y_i, x_j) & \cos(z_i, x_j) \\ \cos(x_i, y_j) & \cos(y_i, y_j) & \cos(z_i, y_j) \\ \cos(x_i, z_j) & \cos(y_i, z_j) & \cos(z_i, z_j) \end{Vmatrix} \qquad (2.2.2)$$

$\mathbf{r} = \overrightarrow{O_i A}$ and $\boldsymbol{\rho} = \overrightarrow{O_j A}$ be radii-vectors of some arbitrary point A with respect to points O_i and O_j, respectively. The matrix T_{ij} is orthogonal, i.e., $T_{ij} T_{ij}^T = T_{ij}^T T_{ij} = E$, where E is a unit matrix. Coordinates of a vector $\mathbf{a}$ with respect to reference frames $O_i x_i y_i z_i$ and $O_j x_j y_j z_j$ are connected by

$$\{\mathbf{a}\}_j = T_{ij} \{\mathbf{a}\}_i, \quad \{\mathbf{a}\}_i = T_{ij}^T \{\mathbf{a}\}_j \qquad (2.2.3)$$

It can be seen from Figure 2.2.2 that

$$\mathbf{r} = \mathbf{R} + \boldsymbol{\rho} \qquad (2.2.4)$$

Let coordinates of the point A in the reference frame $O_j x_j y_j z_j$, as well as coordinates of the point O_j in the frame $O_i x_i y_i z_i$ be known. According to Equations 2.2.3 and 2.2.4, we have the equality

$$\{\mathbf{r}\}_i = \{\mathbf{R}\}_i + T_{ij}^T \{\boldsymbol{\rho}\}_j \qquad (2.2.5)$$

connecting coordinates of the point A in reference frames $O_i x_i y_i z_i$ and $O_j x_j y_j z_j$.

Differentiation of Equation 2.2.5 gives

$$\frac{d}{dt}\{\mathbf{r}\}_i = \frac{d}{dt}\{\mathbf{R}\}_i + \dot{T}_{ij}^{\mathrm{T}}\{\boldsymbol{\rho}\}_j + T_{ij}^{\mathrm{T}}\frac{d}{dt}\{\boldsymbol{\rho}\}_j \qquad (2.2.6)$$

Taking account of Equations 2.2.1 and 2.2.3, we can write Equation 2.2.6 as follows:

$$\{\dot{\mathbf{r}}^{(i)}\}_i = \{\dot{\mathbf{R}}^{(i)}\}_i + \dot{T}_{ij}^{\mathrm{T}}T_{ij}\{\boldsymbol{\rho}\}_i + \{\dot{\boldsymbol{\rho}}^{(i)}\}_i \qquad (2.2.7)$$

Equation 2.2.7 can be replaced by the vector relationship

$$\dot{\mathbf{r}}^{(i)} = \dot{\mathbf{R}}^{(i)} + \dot{\boldsymbol{\rho}}^{(i)} + \Omega\boldsymbol{\rho} \qquad (2.2.8)$$

which connects the velocities of the point A with respect to the reference frames $O_i x_i y_i z_i$ and $O_j x_j y_j z_j$. Here, Ω is a linear operator whose matrix $[\Omega]_i$ in the reference frame $O_i x_i y_i z_i$ is given by (see Equation 2.2.7)

$$[\Omega]_i = \dot{T}_{ij}^{\mathrm{T}}T_{ij} \qquad (2.2.9)$$

We can represent Equation 2.2.8 in an arbitrary reference frame $O_s x_s y_s z_s$. Let us multiply Equation 2.2.7 by the matrix T_{is} from the left, representing $\{\boldsymbol{\rho}\}_i$ as $T_{is}^{\mathrm{T}}\{\boldsymbol{\rho}\}_s$. Taking into account Equations 2.2.3 and 2.2.9, we obtain the desired equality:

$$\{\dot{\mathbf{r}}^{(i)}\}_s = \{\dot{\mathbf{R}}^{(i)}\}_s + T_{is}[\Omega]_i T_{is}^{\mathrm{T}}\{\boldsymbol{\rho}\}_s + \{\dot{\boldsymbol{\rho}}^{(i)}\}_s \qquad (2.2.10)$$

Comparing Equation 2.2.8 with Equation 2.2.10 we arrive at the conclusion that the operator Ω in the reference frame $O_s x_s y_s z_s$ is given by

$$[\Omega]_s = T_{is}[\Omega]_i T_{is}^{\mathrm{T}} \qquad (2.2.11)$$

Let us substitute the relationship $T_{is} = T_{ks}T_{ik}$ into Equation 2.2.11. We obtain

$$[\Omega]_s = T_{ks}T_{ik}[\Omega]_i T_{ik}^{\mathrm{T}}T_{ks}^{\mathrm{T}}$$

Using Equation 2.2.11 with s replaced by k, we come to the formula

$$[\Omega]_s = T_{ks}[\Omega]_k T_{ks}^{\mathrm{T}} \qquad (2.2.12)$$

for transformation of Ω between two arbitrary reference frames $O_s x_s y_s z_s$ and $O_k x_k y_k z_k$. Matrix $[\Omega]_i$ is skew-symmetric, i.e.,

$$[\Omega]_i^{\mathrm{T}} = -[\Omega]_i \qquad (2.2.13)$$

what follows from Equation 2.2.9 and differentiation of the orthogonality condition $T_{ij}^T T_{ij} = E$. It stems from Equations 2.1.12 and 2.1.13 that the matrix of the operator Ω is skew-symmetric in an arbitrary reference frame $O_s x_s y_s z_s$ and, therefore, can be represented as follows:

$$[\Omega]_s = \begin{Vmatrix} 0 & -r^{(s)} & q^{(s)} \\ r^{(s)} & 0 & -p^{(s)} \\ -q^{(s)} & p^{(s)} & 0 \end{Vmatrix} \tag{2.2.14}$$

It is known that the product of skew-symmetric matrix in Equation 2.2.14 and a column $\{a\}_s$ composed of coordinates of an arbitrary vector $\mathbf{a}$ can be represented as a vector (cross) product:

$$[\Omega]_s \{a\}_s = \{\boldsymbol{\omega} \times \mathbf{a}\}_s$$
$$\{\boldsymbol{\omega}\}_s = \|p^{(s)},\ q^{(s)},\ r^{(s)}\|^T \tag{2.2.15}$$

Taking account of Equation 2.2.15, we can write Equation 2.2.8 in the conventional form,

$$\dot{\mathbf{r}}^{(i)} = \dot{\mathbf{R}}^{(i)} + \dot{\boldsymbol{\rho}}^{(j)} + \boldsymbol{\omega} \times \boldsymbol{\rho} \tag{2.2.16}$$

where the vector $\boldsymbol{\omega}$ is the angular velocity of rotation of the reference frame $O_j x_j y_j z_j$ with respect to the reference frame $O_i x_i y_i z_i$. Therefore, $p^{(s)}$, $q^{(s)}$, and $r^{(s)}$ in Equation 2.2.14 are projections of the angular velocity $\boldsymbol{\omega}$ onto the axes of the reference frame $O_s x_s y_s z_s$.

Let us describe the procedure of the solution of Problem 2.3 for a robot with rigid links. Consider an arbitrary point A of the jth link of the manipulator and denote by $\mathbf{r}_s^A = \overrightarrow{O_s A}$ its radius-vector with respect to the origin O_s of the reference frame $O_s x_s y_s z_s$. Assume that the coordinates of the point A with respect to the jth reference frame are known. According to notation adopted in this section, these coordinates are denoted by $\{\mathbf{r}_j^A\}_j$. We will find the coordinates $\{\mathbf{r}_i^A\}_i$ of A with respect to another coordinate frame $O_i x_i y_i z_i$ in terms of $\{\mathbf{r}_j^A\}_j$ and generalized coordinates q.

Let us apply the general Equation 2.2.5 to two adjacent links, distinguishing the cases $i < j$ and $i > j$. Denote $\mathbf{L}_k = O_k O_{k+1}$, $k = 0, 1, \ldots,$ $n - 1$. At first we substitute $j = s$, $i = s - 1$, $\mathbf{r} = \mathbf{r}_{s-1}^A$, $\boldsymbol{\rho} = \mathbf{r}_s^A$, $\mathbf{R} = \mathbf{L}_{s-1}$ into Equation 2.2.5. Then we obtain

$$\{\mathbf{r}_{s-1}^A\}_{s-1} = \{\mathbf{L}_{s-1}\}_{s-1} + T_{s-1,s}^T \{\mathbf{r}_s^A\}_s \tag{2.2.17}$$

Now we take in Equation 2.2.5,

$$j = s,\ i = s + 1,\ \mathbf{r} = \mathbf{r}_{s+1}^A,\ \boldsymbol{\rho} = \mathbf{r}_s^A,\ \mathbf{R} = -\mathbf{L}_s$$

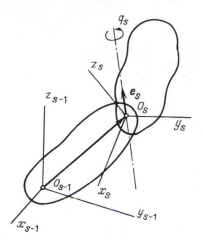

FIGURE 2.2.3.

and get

$$\{\mathbf{r}^A_{s+1}\}_{s+1} = -\{\mathbf{L}_s\}_{s+1} + T^T_{s+1,s}\{\mathbf{r}^A_s\}_s$$

Applying Equation 2.2.3 to the vector $\mathbf{L}_s$, we obtain

$$\{\mathbf{r}^A_{s+1}\}_{s+1} = T_{s,s+1}\left(\{\mathbf{r}^A_s\}_s - \{\mathbf{L}_s\}_s\right) \qquad (2.2.18)$$

Using the recursive Equation 2.2.17 successively for $s = j, j - 1, \ldots ,$ $i + 1$ if $i < j$, and Equation 2.2.18 for $s = j, j + 1, \ldots , i - 1$ if $i > j$, we can express the coordinates $\{\mathbf{r}^A_i\}_i$ through $\{\mathbf{r}^A_j\}_j$, matrices $T_{k,k+1}$ and coordinates $\{\mathbf{L}_k\}_k$. Now we will present these matrices and vectors as functions of the generalized coordinates, considering revolute and prismatic joints separately.

Consider first the revolute joint connecting two adjacent links, the $(s - 1)$th and sth ones (Figure 2.2.3). The generalized coordinate describing the relative position of these links is the angle q_s of rotation of the sth link (with respect to the $(s - 1)$th one) about the axis $\mathbf{e}_s$ of the joint. Let us fix some initial position of the sth link with respect to the $(s - 1)$th one and assume $q_s = 0$ for this position. Let us introduce the auxiliary reference frame $O^0_s x^0_s y^0_s z^0_s$ coinciding with the position of the frame $O_s x_s y_s z_s$ for $q_s = 0$. The matrix $A_s(q_s)$ of direction cosines for the axes of the reference frame $O^0_s x^0_s y^0_s z^0_s$ with respect to the frame $O_s x_s y_s z_s$ is given by[121]

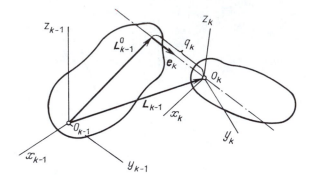

FIGURE 2.2.4.

$$
A_s(q_s) = \left\| \begin{matrix} \cos(x_s^0, x_s) & \cos(y_s^0, x_s) & \cos(z_s^0, x_s) \\ \cos(x_s^0, y_s) & \cos(y_s^0, y_s) & \cos(z_s^0, y_s) \\ \cos(x_s^0, z_s) & \cos(y_s^0, z_s) & \cos(z_s^0, z_s) \end{matrix} \right\|
$$

$$
= \left\| \begin{matrix} (1 - \cos q_s)l_s^2 & (1 - \cos q_s)m_s l_s & (1 - \cos q_s)n_s l \\ + \cos q_s & + n_s \sin q_s & - m_s \sin q_s \\ (1 - \cos q_s)l_s m_s & (1 - \cos q_s)m_s^2 & (1 - \cos q_s)m_s n_s \\ - n_s \sin q_s & + \cos q_s & + l_s \sin q_s \\ (1 - \cos q_s)l_s n_s & (1 - \cos q_s)m_s n_s & (1 - \cos q_s)n_s^2 \\ + m_s \sin q_s & - l_s \sin q_s & + \cos q_s \end{matrix} \right\| \quad (2.2.19)
$$

Here, l_s, m_s, n_s are coordinates of the unit vector e_s of the joint axis in the reference frame $O_s x_s y_s z_s$: $\{e_s\}_s = \|l_s, m_s, n_s\|^T$. Note that l_s, m_s, n_s are known constant parameters. The matrix of direction cosines $T_{s-1,s}(q_s)$ (see Equation 2.2.2) depending on the angle q_s is given by

$$
T_{s-1,s}(q_s) = A_s(q_s)T_{s-1,s}(0) \quad (2.2.20)
$$

The matrix $T_{s-1,s}(0)$ in Equation 2.2.20 corresponds to the initial ($q_s = 0$) position of the sth link with respect to the ($s - 1$)th link. This constant matrix depends only on the choice of reference frames.

Since the sth and ($s - 1$)th links are connected by a revolute joint, the components of the vector $\mathbf{L}_{s-1} = \overline{O_{s-1}O_s}$ are constant in the reference frame $O_{s-1}x_{s-1}y_{s-1}z_{s-1}$: $\{\mathbf{L}_{s-1}\}_{s-1}$ = const. When solving Problem 2.3 we regard the matrix $T_{s-1,s}(0)$ and coordinates $\{\mathbf{L}_{s-1}\}_{s-1}$ as known.

Consider now two links (the ($k - 1$)th and kth ones) connected by a prismatic joint (Figure 2.2.4). In this case the matrix $T_{k-1,k}$ is constant and determined only by the choice of the initial orientation of the reference frames connected with the ($k - 1$)th and kth links. The unit vector e_k of the prismatic joint axis is constant in both these frames, i.e., $\{e_k\}_{k-1}$ = const, $\{e_k\}_k = T_{k-1,k}\{e_k\}_{k-1}$ = const.

Unlike a revolute joint, for a prismatic joint the vector $\mathbf{L}_{k-1} = \overline{O_{k-1}O_k}$ is not constant and obeys the following relationship:

$$\{\mathbf{L}_{k-1}\}_{k-1} = \{\mathbf{L}_{k-1}^0\}_{k-1} + \{\mathbf{e}_k\}_{k-1}q_k \tag{2.2.21}$$

Here, $\mathbf{L}_{k-1}^0$ is the radius-vector of the point O_k with respect to the point O_{k-1} for some initial relative position of the $(k-1)$th and kth links (see Figure 2.2.4); q_k is a linear displacement of the kth link with respect to $(k-1)$th link in the direction of the prismatic joint axis.

As we have mentioned after Equations 2.2.17 and 2.2.18, the coordinates $\{\mathbf{r}_i^A\}_i$ of the given point A with respect to the reference frame $O_i x_i y_i z_i$ can be expressed through the coordinates $\{\mathbf{r}_j^A\}_j$ of this point with respect to the frame $O_j x_j y_j z_j$, matrices $T_{k,k+1}$ and coordinates $\{\mathbf{L}_k\}_k$. The desired expressions are obtained by recursive use of Equations 2.2.17 and 2.2.18. The coordinates $\{\mathbf{r}_j^A\}_j$ are assumed to be known. For a revolute joint, the corresponding matrix $T_{s-1,s}$ is given by Equations 2.2.20 and 2.2.19 as a function of the angle q_s while the coordinates $\{\mathbf{L}_{s-1}\}_{s-1}$ are constant. For a prismatic joint, vice versa, the matrix $T_{k-1,k}$ is constant, and the coordinates $\{\mathbf{L}_{k-1}\}_{k-1}$ depend on the linear displacement q_k (see Equation 2.2.21).

Thus, Equations 2.2.17 through 2.2.21 make it possible to calculate Cartesian coordinates $\{\mathbf{r}_i^A\}_i$ of the point A in terms of the known coordinates $\{\mathbf{r}_j^A\}_j$ and generalized coordinates q_s. Here, indices s lie within the limits $i + 1 \leq s \leq j$ for $i < j$ and $j + 1 \leq s \leq i$ for $i > j$.

Thus, the procedure described above makes it possible to obtain the solution of Problem 2.3 in the form

$$\{\mathbf{r}_i^A\}_i = F(q) \tag{2.2.22}$$

Differentiating Equation 2.2.22 with respect to time, we can obtain formulas for components of the velocity and acceleration of the point A with respect to the reference frame $O_i x_i y_i z_i$. The velocity will be expressed through the generalized coordinates q and velocities $\dot{q}$, and the acceleration will be presented in terms of q, $\dot{q}$, and $\ddot{q}$.

In this section we considered only the general principles of robot kinematics and gave relationships which will be used in subsequent chapters. Note that other approaches, for example, those based on projective geometry and screw calculus, as well as other aspects of robot kinematics, are presented in References 15, 79, 87, 186, 193, 244, 246, 254. These books also contain specific formulas for manipulators of various kinematic structure.

2.3. EQUATIONS OF ROBOT MOTION

2.3.1. DYNAMICS OF SYSTEMS WITH CONSTRAINTS

For a manipulator with rigid links, the radius vector $\mathbf{r}_\nu$ of an arbitrary point A_ν with respect to the inertial reference frame can be expressed through generalized coordinates q (see Equation 2.2.22):

$$\mathbf{r}_\nu = \mathbf{f}_\nu(q), \quad q = (q_1, \ldots, q_n) \tag{2.3.1}$$

Equation 2.3.1 is typical for any mechanical system subject to holonomic constraints. In this connection, let us recall some fundamental concepts of classical mechanics.

Consider first a finite set of particles whose radii-vectors with respect to the origin of some inertial reference frame are $\mathbf{r}_i$, $i = 1, \ldots, k$. Denote Cartesian coordinates of the ith particle in this reference frame by $x_{3i-2}, x_{3i-1}, x_{3i}$. Thus, the total number of coordinates for all particles is $m = 3k$. If possible positions of the particles are not restrained, the system is free and all coordinates x_α, $\alpha = 1, \ldots, 3k$ are independent. In the general case, the motion of the particles can be restricted by constraints:

$$g_s(x, \dot{x}, t) = g_s(x_1, \ldots, x_m, \dot{x}_1, \ldots, \dot{x}_m, t) = 0 \; s = 1, \ldots, p < m \quad (2.3.2)$$

Systems subject to Equation 2.3.2 are called constrained mechanical systems. The constraint is called holonomic if the corresponding function g_s in Equation 2.3.2 is independent of $\dot{x}$, or can be reduced to such a form by integration. Otherwise, the constraint is called nonholonomic. The set of free particles and constrained mechanical systems with all constraints holonomic, are called holonomic mechanical systems. Henceforth, we consider only holonomic systems.

For such a system, we have

$$g_s(x, t) = g_s(x_1, \ldots, x_m, t) = 0, \; s = 1, \ldots, p < m \quad (2.3.3)$$

Therefore, the coordinates $x_1, \ldots, x_m$ are not independent. In the general case, we can express p coordinates, say $x_1, \ldots, x_p$, in terms of the rest coordinates $x_{p+1}, \ldots, x_m$ and time t. Thus, we obtain from Equations 2.3.3

$$x_\beta = x_\beta (x_{p+1}, \ldots, x_m, t), \quad \beta = 1, \ldots, p \quad (2.3.4)$$

The number $n = m - p$ of independent coordinates $x_{p+1}, \ldots, x_m$ is called the number of degrees of freedom for the constrained mechanical system.

Instead of independent Cartesian coordinates $x_{p+1}, \ldots, x_m$ we can introduce a set of n independent variables $q = (q_1, \ldots, q_n)$ such that

$$x_\gamma = h_\gamma(q, t) = h_\gamma(q_1, \ldots, q_n, t), \; \gamma = p + 1, \ldots, m, \; n = m - p \quad (2.3.5)$$

Substituting Equations 2.3.5 into Equations 2.3.4 and combining the resultant equalities with Equations 2.3.5, we obtain

$$x_\alpha = f_\alpha(q, t), \quad \alpha = 1, \ldots, m = 3k \quad (2.3.6)$$

Being presented in the vector form, these equalities become

$$\mathbf{r}_i = \mathbf{f}_i(q, t), \quad i = 1, \ldots, k \quad (2.3.7)$$

The variables $q = (q_1, \ldots, q_n)$ are called generalized coordinates; the number of them is equal to the number of degrees of freedom of the system. If the right-hand sides of Equations 2.3.7 are time independent, the system is called *scleronomous*.

Thus, for holonomic constrained mechanical systems consisting of a finite number of particles, the coordinates of all particles can be expressed in terms of generalized coordinates. Vice versa, solving n Equations 2.3.6 with respect to q and substituting the resultant expressions into the remainder $m - n$ Equations 2.3.6, we obtain $m - n = p$ constraints in the form of Equation 2.3.3. Hence, Equations 2.3.6 or 2.3.7 can be used for the alternative definition of a holonomic mechanical system: the system is called a holonomic system with n degrees of freedom if there exists a set of n independent variables (generalized coordinates) $q = (q_1, \ldots, q_n)$ such that coordinates of all points of the system can be expressed in the form of Equations 2.3.6 or 2.3.7. This definition is suitable not only for a finite number of particles, but also for systems containing continuous bodies, in particular rigid bodies. In this case, the indices α in Equation 2.3.6 and i in Equation 2.3.7 run over a continuous set.

Let us define the concept of virtual displacements. Differentiating Equation 2.3.7 we obtain

$$d\mathbf{r}_i = \sum_{j=1}^{n} \frac{\partial \mathbf{f}_i}{\partial q_j} \, dq_j + \frac{\partial \mathbf{f}_i}{\partial t} \, dt \tag{2.3.8}$$

This relationship describes possible infinitesimal displacements of the ith point of the system, which satisfy the imposed constraints and correspond to increments dq_j of the generalized coordinates q_j, $j = 1, \ldots, n$, and the time interval dt. Vectors $d\mathbf{r}_i$ in Equation 2.3.8 are called *elementary displacements*. The vectors defined by

$$\delta\mathbf{r}_i = \sum_{j=1}^{n} \frac{\partial \mathbf{f}_i}{\partial q_j} \, \delta q_j \tag{2.3.9}$$

are called *virtual displacements*. In Equations 2.3.9 we denote by δq_j increments (variations) of generalized coordinates q_j. The right-hand side of Equation 2.3.9 is the differential of Equation 2.3.7 if time t is regarded as constant. Hence, virtual displacements are elementary displacements under "frozen" constraints corresponding to a fixed instant t. It is evident that elementary and virtual displacements coincide for scleronomous systems.

In classical mechanics, the forces acting on the system are conventionally divided into active forces $\mathbf{F}_i$ and constraint forces $\mathbf{R}_i$. The constraints imposed on the system are called ideal if the work of constraint forces on any virtual displacements is equal to zero, i.e.,

$$\sum_i (\mathbf{R}_i, \delta\mathbf{r}_i) = 0 \qquad\qquad (2.3.10)$$

Here and below, $(\mathbf{a}, \mathbf{b})$ denotes the scalar (dot) product of vectors $\mathbf{a}$ and $\mathbf{b}$. The sum in Equation 2.3.10 is taken over all points of the system. For continuous bodies, such sums should be replaced by integrals, taken over the body volume.

Usually, manipulation robots are treated as mechanical systems with ideal constraints. Those constraint forces that violate Equation 2.3.10 (e.g., friction forces) can be regarded as active forces.

The most widespread form of equations for holonomic mechanical systems with ideal constraints are the Lagrange equations:

$$\frac{d}{dt}\frac{\partial T}{\partial \dot{q}_j} - \frac{\partial T}{\partial q_j} = Q_j, \quad j = 1, \ldots, n \qquad\qquad (2.3.11)$$

$$T = \frac{1}{2}\sum_i m_i\dot{\mathbf{r}}_i^2 \qquad\qquad (2.3.12)$$

$$Q_j = \sum_i \left(\mathbf{F}_i, \frac{\partial \mathbf{r}_i}{\partial q_j}\right) = \sum_i \left(\mathbf{F}_i, \frac{\partial \mathbf{f}_i}{\partial q_j}\right) \qquad\qquad (2.3.13)$$

Here, T is the kinetic energy of the system; $\mathbf{F}_i$ is the vector sum of active forces applied to the ith point of the system; Q_j is the generalized force corresponding to the generalized coordinate q_j; m_i is the mass of the ith particle of the system; and vector functions $\mathbf{f}_i$ are defined by Equation 2.3.7. The dot in Equation 2.3.12 denotes the time derivative of the vector $\mathbf{r}_i$ with respect to the inertial reference frame. The sums in Equations 2.3.12 and 2.3.13 are taken over all particles of the system.

In robotics, the active forces $\mathbf{F}_i$ include control forces created by actuators, different external forces such as gravity forces, drag, friction in joints, forces of interaction of the manipulator with other objects (workpieces, tools), etc.

The generalized forces Q_j can be obtained from Equation 2.3.13 if the active forces $\mathbf{F}_i$ and kinematic relationships (Equations 2.3.7) are known. However, another procedure, based on the expression for the work produced by active forces on virtual displacements, seems to be more rational.

The work δA of active forces applied to the particles of the system on arbitrary virtual displacements $\delta \mathbf{r}_i$ is defined by

$$\delta A = \sum_i (\mathbf{F}_i, \delta \mathbf{r}_i)$$

Using Equations 2.3.9 and 2.3.13 we obtain the following expression for the work δA:

$$\delta A = \sum_i \left(\mathbf{F}_i, \sum_{j=1}^n \frac{\partial \mathbf{f}_i}{\partial q_j} \delta q_j \right) = \sum_{j=1}^n \sum_i \left(\mathbf{F}_i, \frac{\partial \mathbf{f}_i}{\partial q_j} \right) \delta q_j = \sum_{j=1}^n Q_j \delta q_j \quad (2.3.14)$$

Since the generalized coordinates $q_1, \ldots, q_n$ are independent, Equation 2.3.14 implies the simple rule for calculating generalized forces. To find the force Q_j we determine the work produced by all active forces on virtual displacements corresponding to the increment δq_j of the respective coordinate q_j, other generalized coordinates being fixed. Then we obtain from Equation 2.3.14: $Q_j = \delta A / \delta q_j$.

2.3.2. CALCULATION OF KINETIC ENERGY

To compose Lagrange Equations 2.3.11, it is necessary to express kinetic energy (Equation 2.3.12) in terms of generalized coordinates q, velocities $\dot{q}$, and time t. The basic relationship used for calculating the kinetic energy of a manipulation robot with rigid links is the formula for the kinetic energy of a rigid body.

Let us introduce the denotations, consistent with those of the previous section: $O_0 x_0 y_0 z_0$ is the inertial reference frame; $O_s x_s y_s z_s$ is the reference frame connected with the rigid body, for example, with an arbitrary sth link of a manipulator; $\mathbf{r}_s^0 = \overrightarrow{O_0 O_s}$; $\mathbf{r}_s^C = \overrightarrow{O_s C_s}$ is the radius-vector of the center of mass (C_s) of the body with respect to the point O_s; $\boldsymbol{\omega}_s$ is the angular velocity of the body with respect to the inertial reference frame; M_s is the mass of the body; J_s is its tensor of inertia with respect to the point O_s. The kinetic energy T_s of the rigid body is determined by

$$T_s = \frac{1}{2} M_s (\dot{\mathbf{r}}_s^0)^2 + \frac{1}{2}(J\boldsymbol{\omega}_s, \boldsymbol{\omega}_s) + M_s(\dot{\mathbf{r}}_s^0, \dot{\mathbf{r}}_s^C) \quad (2.3.15)$$

Inertia properties of a rigid body are completely determined by its mass M_s and the tensor of inertia J_s with respect to the point O_s. The tensor of inertia J_s is a symmetric tensor of the second order. In an arbitrary Cartesian reference frame $O_s xyz$ with the origin O_s, the tensor J_s is represented by a symmetric matrix:

$$[J_s] = \begin{Vmatrix} J_{11} & J_{12} & J_{13} \\ J_{21} & J_{22} & J_{23} \\ J_{31} & J_{32} & J_{33} \end{Vmatrix}$$

$$J_{11} = \sum_i m_i(y_i^2 + z_i^2), \quad J_{22} = \sum_i m_i (x_i^2 + z_i^2)$$

$$J_{33} = \sum_i m_i (x_i^2 + y_i^2)$$

$$J_{12} = J_{21} = -\sum_i m_i x_i y_i, \quad J_{13} = J_{31} = -\sum_i m_i x_i z_i$$

$$J_{23} = J_{32} = -\sum_i m_i y_i z_i \qquad (2.3.16)$$

Here, x_i, y_i, and z_i are coordinates of the
ith particle of the body with respect to the frame $O_s xyz$, and m_i is the mass
of the particle.

Diagonal elements J_{11}, J_{22}, and J_{33} of the matrix $[J_s]$ are the moments of
inertia of the body about axes $O_s x$, $O_s y$, and $O_s z$, respectively. Off-diagonal
elements of the matrix $[J_s]$ are called products of inertia.

For a compact presentation of Equation 2.3.16, it is convenient to use
the index notation. Denote by ρ_i the radius-vector of the ith point of a body
with respect to O_s, and by ρ_i^1, ρ_i^2, and ρ_i^3 its coordinates in the reference frame
$O_s xyz$. Then the general expression for an element $J_{\alpha\beta}$, $\alpha, \beta = 1, 2, 3$, of
the matrix $[J_s]$ can be written as

$$J_{\alpha\beta} = \sum_i m_i (\rho_i^2 \delta_{\alpha\beta} - \rho_i^\alpha \rho_i^\beta), \quad \alpha, \beta = 1, 2, 3$$

$$\delta_{\alpha\beta} = \begin{cases} 1, & \text{if } \alpha = \beta \\ 0, & \text{if } \alpha \neq \beta \end{cases}$$

When changing the reference frame $O_s xyz$ for the frame $O_s x'y'z'$ with different
orientation, the matrix of the tensor of inertia is transformed as follows:

$$[J_s]' = A[J_s]A^T, \quad [J_s] = A^T[J_s]'A \qquad (2.3.17)$$

Here, A is the corresponding transformation given by Equation 2.2.2 in which
one should substitute

$$x_i = x, \quad y_i = y, \quad z_i = z, \quad x_j = x', \quad y_j = y', \quad z_j = z'$$

There exists the coordinate frame $O_s x_p y_p z_p$ such that the products of inertia
of the body are equal to zero in this frame, and the tensor of inertia J_s is
diagonal:

$$[J_s]_p = \begin{Vmatrix} J_{11} & 0 & 0 \\ 0 & J_{22} & 0 \\ 0 & 0 & J_{33} \end{Vmatrix} = \text{diag } (J_{11}, J_{22}, J_{33})$$

Coordinate axes of the frame $O_s x_p y_p z_p$ are called principal axes of inertia of the body at the point O_s. Moments of inertia $J_{\alpha\alpha}$, $\alpha = 1, 2, 3$, about the axes $O_s x_p$, $O_s y_p$, and $O_s z_p$ are called principal moments of inertia of the body at the point O_s.

When proceeding from the reference frame $O_s xyz$ to the frame $C_s xyz$ with the origin C_s at the center of mass of the body and axes parallel to the corresponding axes of the frame $O_s xyz$, the components of the tensor of inertia are transformed as follows:

$$J_{\alpha\beta}^C = J_{\alpha\beta} - M_s \, [(\mathbf{r}_s^C)^2 \delta_{\alpha\beta} - (\mathbf{r}_s^C)^\alpha (\mathbf{r}_s^C)^\beta], \quad \alpha, \beta = 1, 2, 3 \quad (2.3.18)$$

Here, $J_{\alpha\beta}^C$, $\alpha, \beta = 1, 2, 3$, are the components of the tensor of inertia of the body at its center of mass while $(\mathbf{r}_s^C)^\alpha$ are components of the vector $\mathbf{r}_s^C = \overrightarrow{O_s C_s}$.

Using the tensor of inertia J_s, we can find the moment of inertia I_e of the rigid body about any axis passing through the point O_s:

$$I_e = (J_s \mathbf{e}, \mathbf{e}) \tag{2.3.19}$$

Here, $\mathbf{e}$ is the unit vector of the axis.

Applying the rule given by Equation 2.3.19 to both sides of Equations 2.3.18 we obtain the well-known Huygens-Steiner theorem:

$$I_e = I_e^C + M_s d^2$$

Here, I_s^C is the moment of inertia about the axis parallel to $\mathbf{e}$ and passing through the center of mass C_s of the body, d is the distance between the parallel axes passing through the points O_s and C_s.

Having determined the tensor of inertia with respect to some reference frame, we can obtain this tensor with respect to any other reference frame using Equations 2.3.17 and 2.3.18 for transformations of rotation and translation, respectively.

Concrete formulas for moments of inertia for various bodies can be found in books on classical mechanics and relevant handbooks. If the shape of the body and the mass distribution are known, the calculation of moments of inertia is reduced to integration over the volume of the body. Sometimes the position of the center of inertia of the body and its moments of inertia are obtained experimentally.

The kinetic energy of the manipulator is the sum of kinetic energies of its parts

$$T = \sum_s T_s \tag{2.3.20}$$

Each term of Equation 2.3.20 has the form of Equation 2.3.15. To obtain the Lagrange equations, we are to express the components of the vectors $\dot{\mathbf{r}}_s^0$, $\dot{\mathbf{r}}_s^C$, $\boldsymbol{\omega}_s$ in terms of generalized coordinates and velocities. It can be done using the procedure described in Section 2.2.

By successive application of recursive Equation 2.2.17, we can express the coordinates $\{\mathbf{r}_s^0\}_0$ of the point O_s in the inertial reference frame $O_0x_0y_0z_0$ through the generalized coordinates of the manipulator (see Equation 2.2.22):

$$\{\mathbf{r}_s^0\}_0 = F_s(q_1, \ldots, q_{s-1}) \tag{2.3.21}$$

Differentiating Equation 2.3.21 with respect to time we obtain the components of the velocity $\dot{\mathbf{r}}_s^0$ of the point O_s with respect to the inertial reference frame:

$$\{\dot{\mathbf{r}}_s^0\}_0 = \sum_{k=1}^{s-1} \frac{\partial F_s}{\partial q_k} \dot{q}_k \tag{2.3.22}$$

Applying the second Equation 2.2.3 to the vector $\mathbf{r}_s^C = \overrightarrow{O_sC_s}$ with $i = 0, j = s$, we get

$$\{\mathbf{r}_s^C\}_0 = T_{0s}^T (q) \{\mathbf{r}_s^C\}_s$$

$$T_{0s}(q_1, \ldots, q_s) = T_{s-1,s}(q_s)T_{s-2,s-1}(q_{s-1}) \ldots T_{01}(q_1) \tag{2.3.23}$$

Here, matrices $T_{k-1,k}$, $k = 1, \ldots, s$ are given by Equations 2.2.19 and 2.2.20 for a revolute joint and are constant matrices for prismatic joints.

Note that coordinates $\{\mathbf{r}_s^C\}_s$ are constant, since the center of mass of the *s*th link is fixed with respect to this link. Differentiating Equation 2.3.23 we obtain

$$\{\dot{\mathbf{r}}_s^C\}_0 = \sum_{k=1}^{s} \left[\frac{\partial T_{0s}^T}{\partial q_k} \dot{q}_k \right] \{\mathbf{r}_s^C\} \tag{2.3.24}$$

For calculating kinetic energy (Equation 2.3.15), it seems to be convenient to find the projections $p^{(s)}$, $q^{(s)}$, and $r^{(s)}$ of the absolute angular velocity $\boldsymbol{\omega}_s$ of the *s*th link onto the axes of the frame $O_sx_sy_sz_s$ connected with the link. Using Equations 2.2.9, 2.2.11 with $i = 0$ and $j = s$, 2.3.23, and 2.2.14 we arrive at

$$[\Omega]_s = T_{0s}\dot{T}_{0s}^{\mathrm{T}} = \sum_{k=1}^{s} T_{0s}\frac{\partial T_{0s}^{\mathrm{T}}}{\partial q_k}\dot{q}_k$$

$$p^{(s)} = \Omega_{32}, \quad q^{(s)} = \Omega_{13}, \quad r^{(s)} = \Omega_{21} \tag{2.3.25}$$

Here, Ω_{13}, Ω_{21}, and Ω_{32} denote corresponding elements of the skew-symmetric matrix $[\Omega]_s$. Note that by Equation 2.3.25 $p^{(s)}$, $q^{(s)}$, and $r^{(s)}$ are expressed in terms of generalized coordinates $q_1, \ldots, q_s$ and velocities $\dot{q}_1, \ldots, \dot{q}_s$.

Substituting expressions given by Equations 2.3.22, 2.3.24, and 2.3.25 into Equation 2.3.15, we obtain the kinetic energy T_s of the sth link of the manipulator as a function of the generalized coordinates and velocities. Since $\{\dot{\mathbf{r}}_s^0\}_0$, $\{\dot{\mathbf{r}}_s^C\}_0$, $p^{(s)}$, $q^{(s)}$, and $r^{(s)}$ depend on $\dot{q}$ linearly, the function T_s is a quadratic form of the corresponding generalized velocities, its coefficients depending on the generalized coordinates. Equation 2.3.20 gives the total kinetic energy of the system as a quadratic form of generalized velocities:

$$T = \frac{1}{2}\sum_{i,j=1}^{n} a_{ij}(q)\,\dot{q}_i\dot{q}_j, \quad a_{ij} = a_{ji} \tag{2.3.26}$$

Note that, apart from the links, manipulators usually contain other moving parts: rotors of electric motors, gears, etc. We can treat them as additional links connected to corresponding arm links by kinematic paris and carry out the above procedure for calculating the kinetic energy. However, generalized coordinates of these parts, say, angles of rotation of rotors, usually are not independent. For example, let q_s be the rotation angle of the sth link with respect to the $(s-1)$th one, and ψ_s be the angle of rotation (with respect to the $(s-1)$th link) of the rotor of the actuator placed at the sth joint. The variables q_s and ψ_s are related by $\psi_s = n_s q_s$, where n_s is a gear ratio of the corresponding reduction gear. In such cases additional moving parts do not increase the number of degrees of freedom of the manipulator mechanical system.

2.3.3. Dynamic Equations

Substituting kinetic energy (Equation 2.3.26) into Lagrange Equations 2.3.11, we obtain

$$\sum_{j=1}^{n} a_{ij}(q)\,\ddot{q}_j + \frac{1}{2}\sum_{j,k=1}^{n}\left[\frac{\partial a_{ij}(q)}{\partial q_k} - \frac{\partial a_{ik}(q)}{\partial q_j} - \frac{\partial a_{jk}(q)}{\partial q_i}\right]\dot{q}_j\dot{q}_k = Q_i \quad i = 1, \ldots, n \tag{2.3.27}$$

Here, the generalized forces Q_i (see Equation 2.3.13) include various external and internal forces, such as gravity forces, friction, drag, electromagnetic torques created by actuators, forces produced by pneumatic and hydraulic drives, etc. The generalized forces may depend on time t, the generalized

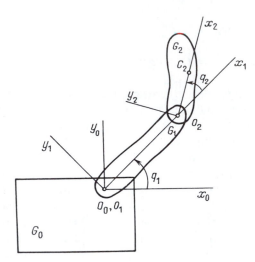

FIGURE 2.3.1.

coordinates q, and velocities $\dot{q}$, as well as on control parameters: electric voltages and/or currents, pneumatic and hydraulic pressures, etc.

The initial conditions for the system described by Equations 2.3.27 are usually given in the conventional form:

$$q_i\,(t_0) = q_i^0, \quad \dot{q}_i\,(t_0) = \dot{q}_i^0, \quad i = 1, \ldots, n \qquad (2.3.28)$$

Here, t_0 is the initial time instant; q_i^0 and $\dot{q}_i^0$ are given constants.

Lagrange Equations 2.3.27 describe the dynamics of the mechanical part of the robot. These equations should be supplemented with equations for physical processes in electric, pneumatic, or hydraulic actuators. Depending on the assumed actuators models, the additional equations can be either finite or differential. In the latter case, the number of state variables increases, and additional initial conditions are required.

2.3.4. EQUATIONS OF MOTION FOR A TWO-LINK MANIPULATOR

Let us illustrate the procedure of deriving equations of motion by an example of a two-link electromechanical robot. We assume that the robot arm moves in a horizontal plane. The top view of the robot is given in Figure 2.3.1. The robot consists of a fixed base G_0, two rigid links G_1 and G_2, and two actuators D_1 and D_2. The link G_1 is connected to the base G_0 and to the link G_2 by revolute joints with vertical axes. Each actuator consists of a DC motor with independent excitation and a reduction gear. The actuator D_1 mounted on the base G_0 controls the rotation of the first link with respect to the base, while the actuator D_2 placed on the link G_1 controls the relative rotation of the links. We suppose that the structure of the actuators is similar to that considered in Section 2.1.2 and shown in Figure 2.1.1. We assume that the actuator rotors rotate about their axes of dynamic symmetry.

Let us connect the right-hand Cartesian reference frames $O_0x_0y_0z_0$, $O_1x_1y_1z_1$, and $O_2x_2y_2z_2$ with the base G_0 and the links G_1 and G_2 as shown in Figure 2.3.1. The axes O_sx_s are pointed vertically. The origin O_2 of the frame $O_2x_2y_2z_2$ is placed at the intersection point of the horizontal plane π passing through the mass center C_2 of the link G_2 with the axis of the joint connecting the links (the second joint). The origins O_0 and O_1 coincide with the intersection point of the plane π with the axis of the first joint (connecting the first link with the base). The axis O_1x_1 passes through the point O_2, and the axis O_2x_2 passes through the mass center C_2 of the second link. The direction of the axis O_0x_0 can be chosen arbitrarily in the plane π.

Let us take the angle q_1 between the axes O_0x_0 and O_1x_1 (the rotation angle of the first link with respect to the base) and the angle q_2 between the axes O_1x_1 and O_2x_2 (the relative rotation angle of the links) as generalized coordinates. Denote by M the mass of the second link; by m the mass of the rotor of the actuator D_2; by I_1 the moment of inertia of the link G_1 (together with the actuator parts rigidly connected with this link) about the axis of the first joint; by I_2 the moment of inertia of the link G_2 about the axis of the second joint; by J_1 and J_2 the inertia moments of the actuator rotors about their axes; by ψ_1 the rotation angle of the first rotor with respect to the base; by ψ_2 the rotation angle of the second rotor with respect to the link G_1; by n_s the gear ratio of the actuator D_s, $s = 1, 2$; by μ_s the electromagnetic torque created by the actuator D_s and applied to its rotor. The angles q_i and ψ_i are connected by

$$\psi_i = n_iq_i, \quad i = 1, 2 \tag{2.3.29}$$

The kinetic energy of the manipulator is the sum of kinetic energies of four rigid bodies: two links and two rotors.

For the first link according to Equation 2.3.15 we obtain

$$T_1 = \frac{1}{2}I_1\dot{q}_1^2, \quad (\dot{\mathbf{r}}^0 = 0, \ \boldsymbol{\omega}_1 = (0, 0, \dot{q}_1)) \tag{2.3.30}$$

For the second link we have

$$\{\dot{\mathbf{r}}_2^0\}_0 = \| - a \sin q_1\dot{q}_1, \quad a \cos q_1\dot{q}_1, \quad 0\|^{\mathrm{T}}$$

$$\{\dot{\mathbf{r}}_2^C\}_0 = \| - L \sin (q_1 + q_2) (\dot{q}_1 + \dot{q}_2),$$

$$L \cos (q_1 + q_2) (\dot{q}_1 + \dot{q}_2), \quad 0\|^{\mathrm{T}}$$

$$\boldsymbol{\omega}_2 = (0, 0, \dot{q}_1 + \dot{q}_2)$$

Here, $a = O_1O_2$, $L = O_2C$. Using equation 2.3.15 for the second link, we get

$$T_2 = \frac{1}{2}(Ma^2 + I_2 + 2MaL \cos q_2)\dot{q}_1^2 + \frac{1}{2}I_2\dot{q}_2^2 +$$

$$(I_2 + MaL \cos q_2)\dot{q}_1\dot{q}_2 \qquad (2.3.31)$$

Kinetic energies T_3, T_4 for the rotors are calculated in a similar way. The expression for T_3 is given by Equation 2.3.30 with I_1 replaced by J_1 and $\dot{q}_1$ replaced by $\dot{\psi}_1 = n_1\dot{q}_1$ (see Equation 2.3.29):

$$T_3 = \frac{1}{2}J_1 n_1^2 \dot{q}_1^2 \qquad (2.3.32)$$

The kinetic energy T_4 can be obtained from Equation 2.3.31 if we formally put $M = m$, $I_2 = J_2$, $L = 0$ (the center of mass of the second rotor lies on its axis), and $\dot{q}_2 = \dot{\psi}_2 = n_2\dot{q}_2$. Thus we get

$$T_4 = \frac{1}{2}(ma^2 + J_2)\dot{q}_1^2 + \frac{1}{2}J_2 n_2^2 \dot{q}_2^2 + J_2 n_2 \dot{q}_1\dot{q}_2 \qquad (2.3.33)$$

Adding Equations 2.3.30 through 2.3.33 we obtain the total kinetic energy of the manipulator:

$$T = T_1 + T_2 + T_3 + T_4 = \frac{1}{2}(A_{11} + 2MaL \cos q_2)\dot{q}_1^2 + \frac{1}{2}A_{22}\dot{q}_2^2$$

$$+ (A_{12} + MaL \cos q_2)\dot{q}_1\dot{q}_2$$

$$A_{11} = (M + m)a^2 + I_1 + I_2 + J_1 n_1^2 + J_2$$

$$A_{12} = I_2 + J_2 n_2, \quad A_{22} = I_2 + J_2 n_2^2 \qquad (2.3.34)$$

Let us determine now the generalized forces Q_1, Q_2, corresponding to the coordinates q_1, q_2. The only active forces applied to the system are the electromagnetic control forces produced by the electric motors. Virtual work (Equation 2.3.14) of these forces is given by

$$\delta A = \mu_1\delta\psi_1 + \mu_2\delta\psi_2 = \mu_1 n_1\delta q_1 + \mu_2 n_2\delta q_2 \qquad (2.3.35)$$

Here, constraints (Equation 2.3.29) are taken into account. It stems from Equation 2.3.35 that the generalized forces Q_1, Q_2 are

$$Q_1 = \mu_1 n_1, \quad Q_2 = \mu_2 n_2 \qquad (2.3.36)$$

Inserting kinetic energy (Equation 2.3.34) and generalized forces (Equation 2.3.36) into Lagrange Equations 2.3.11, we obtain the dynamical equations for our manipulator:

$$(A_{11} + 2MaL \cos q_2)\ddot{q}_1 + (A_{12} + MaL \cos q_2)\ddot{q}_2$$

$$- 2MaL \sin q_2\dot{q}_1\dot{q}_2 - MaL \sin q_2\dot{q}_2^2 = n_1\mu_1$$

$$(A_{12} + MaL \cos q_2)\ddot{q}_1 + A_{22}\ddot{q}_2 + MaL \sin q_2\dot{q}_1^2 = n_2\mu_2 \quad (2.3.37)$$

The driving torques μ_1, μ_2 are created by electric motors and controlled by changing the voltages applied to the armature circuits. Therefore, Equations 2.3.37 should be supplemented by additional equations relating the torques to variables describing processes in electric circuits. For DC motors with independent excitation these additional equations are given by Equations 2.1.9:

$$L_s\frac{dj_s}{dt} + R_sj_s + k_{1s}\dot{\psi}_s = u_s$$

$$\mu_s = k_{2s}j_s, \quad s = 1, 2 \quad (2.3.38)$$

Here, L_s and R_s are the inductance and resistance of the armature winding of the actuator D_s; k_{1s} and k_{2s} are constant parameters of the sth electric motor; j_s is the electric current in the armature circuit of the sth motor; u_s is the input control voltage of the corresponding actuator.

Equations 2.3.37 and 2.3.38 describe the mathematical model of the electromechanical two-link manipulator shown in Figure 2.3.1.

2.3.5. CONCLUDING REMARKS

We presented above one possible approach to describing dynamics of manipulation robots by means of the Lagrange equations. We considered only manipulators with perfectly rigid parts. The Lagrange approach can also be used for robots containing deformable parts: springs, elastic links, etc. In this case mathematical models of manipulators may include partial differential and/or integrodifferential equations.

The Lagrange equations are used in subsequent chapters of this book to describe manipulation robot dynamics.

Other types of equations can also be applied to modeling robot dynamics, for example, the Hamiltonian equations, the Routh equations,[185] the Euler-Lagrange equations using quasi-velocities (linear combinations of generalized velocities),[158] etc.

The equations of motion can also be composed by applying the linear and angular momentum theorems to separate parts of the manipulator. This approach implies the procedure of eliminating constraint forces in kinematic pairs.

For modeling robot dynamics, some authors prefer direct use of variational principles of mechanics. Such an approach does not lead explicitly to differential equations.

In Reference 193, for example, the method based on the Gauss minimum compulsion principle is described.

Thus, there are a number of methods for obtaining equations of robot dynamics. These methods are based on different principles of classical mechanics and presented in detail in many books on robotics; see, for instance, References 15, 87, 144, 186, 193, 241, 246, 247, 254.

For multilink robots, the implementation of any of these methods leads to cumbersome analytical calculations. The resulting equations turn out to be very lengthy. For this reason, methods of computer algebra are widely used for deriving equations of motion for multilink robots.[241]

As a rule, equations of motion of robots are nonlinear and complicated. They are solved by various numerical methods, some of them being specially adjusted for robots. Different aspects of modeling and simulation of robot dynamics are extensively discussed in a number of books on robotics (see, for instance, References 15, 87, 144, 193, 241, 246).

Chapter 3

DYNAMICS OF SYSTEMS WITH ELASTIC AND DISSIPATIVE ELEMENTS

3.1. MECHANICAL SYSTEMS WITH ELASTIC AND DISSIPATIVE ELEMENTS

3.1.1. DESCRIPTION OF MECHANICAL SYSTEMS

In this chapter we consider a special class of mechanical systems containing elastic elements of high stiffness and dissipative elements. Using asymptotic methods of small parameter, we analyze the Lagrange equations for such systems and find the general structure of their solutions for some cases. We determine the additional (as compared to the rigid model) displacements, velocities, and forces due to elastic compliance of the elements. The results obtained here are used in the following chapters for the analysis of dynamics of robots with elastic links and/or joints. In this chapter we present the results first obtained in References 57 and 58.

We consider a holonomic scleronomous mechanical system with $n + m$ degrees of freedom subject to ideal constraints. The system contains elastic elements whose stiffness is high and proportional to ϵ^{-2}, where $\epsilon \ll 1$ is a small parameter. In the limit $\epsilon \to 0$, the elastic elements become rigid and are converted into additional m ideal constraints. Here, the initial system becomes a system with n degrees of freedom; this degenerate system will be called a rigid model.

We denote by $s = (s_1, \ldots, s_n)$ the generalized coordinates of the rigid model, and by $x = (x_1, \ldots, x_m)$ the additional generalized coordinates of the initial system that are due to the elastic elements. The system contains inertial elements (point masses and/or rigid bodies) whose position can be characterized by the N coordinates $q = (q_1, \ldots, q_N)$. The kinematics of the system is specified by the relation

$$q = f(s, x) \tag{3.1.1}$$

where f is a given vector-valued function. We assume that the matrices of partial derivatives

$$F(s, x) = \left\| \frac{\partial f_i}{\partial s_j} \right\|, \qquad \Phi(s, x) = \left\| \frac{\partial f_i}{\partial x_k} \right\|$$

$$i = 1, \ldots, N; \qquad j = 1, \ldots, n; \qquad k = 1, \ldots, m \tag{3.1.2}$$

have maximal possible ranks equal to $r = \min(n, N)$ for the matrix F, and $\rho = \min(m, N)$ for the matrix Φ. We assume also that $N \leq n + m$. If one

65

of these three assumptions is violated we can decrease the dimension of one of the three vectors: q, s, or x. The inequality $N < n + m$ means that some of the degrees of freedom corresponding to coordinates s and x are inertia free, i.e., they correspond to zero masses. We will call the variables s, x, and q as "rigid", "elastic", and "inertial" coordinates, respectively.

The kinetic energy of the system has the form

$$T = \frac{1}{2}(A(q)\dot{q}, \dot{q}) = \frac{1}{2}\dot{q}^T A(q)\dot{q} = \frac{1}{2}\sum_{i,j=1}^{N} a_{ij}(q)\dot{q}_i\dot{q}_j \qquad (3.1.3)$$

Here, $A = \{a_{ij}\}$ is a symmetric positive definite $N \times N$ matrix.

In the case of a rigid model, the system is acted upon by generalized forces $Q^s = \{Q_i^s\}$ that perform work on the displacements δs_i, $i = 1, \ldots, n$, and forces $Q^q = \{Q_i^q\}$ that perform work on the displacements δq_i, $i = 1, \ldots, N$. For the initial system, we add the forces resulting from the potential energy Π of the elastic elements, and also external forces $Q^x = \{Q_i^x\}$, corresponding to the displacements δx_i, $i = 1, \ldots, m$. We assume that elastic forces are linear functions of displacements x, and dissipative forces are linear functions of velocities $\dot{x}$. Therefore, the potential energy and Rayleigh dissipative function are the quadratic forms

$$\Pi = \frac{1}{2}x^T C(s)x, \qquad R = \frac{1}{2}\dot{x}^T B(s)\dot{x} \qquad (3.1.4)$$

Here, C and B are symmetric $m \times m$ matrices, C is positive definite, and B is nonnegative definite.

Let us comment on the forces introduced above. The forces Q^q acting on inertial elements are external forces such as the gravity force, resistance of the medium, etc. They depend on q, $\dot{q}$, and t.

The forces Q^x include forces acting in the elastic and dissipative elements, besides those linear forces corresponding to functions in Equation 3.1.4. The forces Q^x can describe nonlinearity of elastic and dissipative elements.

The forces Q^s include control forces Q^*, depending on time t and also external forces Q'. We assume that

$$Q^q = Q^q(q, \dot{q}, t), \qquad Q^x = Q^x(s, \dot{s}, x, \dot{x}, t)$$

$$Q^* = Q'(s, \dot{s}, x, \dot{x}, t) + Q^*(t) \qquad (3.1.5)$$

Here, Q^q, Q^x, and Q' are specified functions, while the control forces Q^* can be either given or unknown. The fact that forces Q^x and Q' are independent of q and $\dot{q}$ entails no loss of generality, since q is expressed in terms of s and x in accordance with Equation 3.1.1. Note that due to Equation 3.1.1, some external forces can be treated in different ways: either as Q^q or as Q^s and Q^x. This nonuniqueness will not change the ultimate form of equations of motion.

For manipulation robots with elastic elements (joints and/or links), the "rigid" coordinates s correspond to the degrees of freedom of the rigid model, namely, angles of rotation and linear displacements in kinematic pairs. The "elastic" coordinates x are elastic displacements, and the "inertial" coordinates q are the generalized coordinates of the inertial elements (links, payload). If we take into account masses of all elements, then $N = n + m$. However, sometimes it is possible to neglect masses of some elements (links); in this case $N < n + m$. For a manipulator, the forces Q^* are control torques and/or forces created by actuators, forces Q' may include the friction torques in joints, Q^q are external forces (gravity, external resistance), and Q^x may correspond to the nonlinearity of elastic elements.

Our basic simplifying assumption is that the stiffness of the elastic elements is high, namely,

$$C(s) = \epsilon^2 K(s), \qquad \epsilon \ll 1 \tag{3.1.6}$$

Here, K is a symmetric positive definite $m \times m$ matrix, while ϵ is a nonnegative small parameter. Moreover, we make one of the following two assumptions: either the matrix B is bounded as $\epsilon \to 0$, i.e.,

$$B(s) = O(1), \qquad \epsilon \to 0 \tag{3.1.7}$$

or it can be presented as

$$B(s) = \epsilon^{-1} D(s) = O(\epsilon^{-1}) \tag{3.1.8}$$

Here, D is a nonnegative definite $m \times m$ matrix. All the remaining mechanical quantities (forces, masses, and so forth), namely, f, F, Φ, A, K, Q^q, Q^s, Q', Q^*, are assumed to be independent of ϵ. The elastic forces corresponding to the potential energy Π from Equation 3.1.4 and equal to $-\partial\Pi/\partial x_i$, are also bounded as $\epsilon \to 0$. Then it follows from Equation 3.1.6 that the elastic displacements x are small and can be presented as

$$x = \epsilon^2 X, \qquad X = O(1) \tag{3.1.9}$$

where X is bounded as $\epsilon \to 0$.

Therefore, as $\epsilon \to 0$, our system with $n + m$ degrees of freedom is converted into the rigid model with n degrees of freedom, for which $x \equiv 0$.

3.1.2. EQUATIONS OF MOTION

The work δA of all forces on the virtual displacements is given by

$$\delta A = \sum_{i=1}^{n} Q_i^s \delta s_i + \sum_{i=1}^{N} Q_i^q \delta q_i + \sum_{i=1}^{m} \left(Q_i^x - \frac{\partial R}{\partial \dot{x}_i} \right) \delta x_i - \delta\Pi \tag{3.1.10}$$

Here, the dissipative forces are expressed as the partial derivatives of the function R from Equation 3.1.4 with respect to the velocities $\dot{x}_i$. We express the increments δq_i in Equation 3.1.10 in terms of δs_i and δx_i using Equations 3.1.1 and 3.1.2. Thus we get

$$\delta A = \sum_{i=1}^{n} \left[Q_i^s - \frac{\partial \Pi}{\partial s_i} + (F^T Q^q)_i \right] \delta s_i$$

$$+ \sum_{i=1}^{m} \left[Q_i^x - \frac{\partial \Pi}{\partial x_i} - \frac{\partial R}{\partial \dot{x}_i} + (\Phi^T Q^q)_i \right] \delta x_i \qquad (3.1.11)$$

We set up the Lagrange equations, taking s and x as generalized coordinates and using Equations 3.1.11 and 3.1.4:

$$\frac{d}{dt}\frac{\partial T}{\partial \dot{s}} - \frac{\partial T}{\partial s} - F^T Q^q = Q^s - \frac{1}{2}\frac{\partial (Cx, x)}{\partial s}$$

$$\frac{d}{dt}\frac{\partial T}{\partial \dot{x}} - \frac{\partial T}{\partial x} - \Phi^T Q^q = Q^x - Cx - B\dot{x} \qquad (3.1.12)$$

Here and henceforth, the derivatives with respect to a vector are considered as the gradients.

We introduce the notation

$$\Gamma(q, \dot{q}, \ddot{q}, t) = \frac{d}{dt}\frac{\partial T}{\partial \dot{q}} - \frac{\partial T}{\partial q} - Q^q \qquad (3.1.13)$$

It can be easily shown using only general relationships (Equation 3.1.1) that the left sides of Equations 3.1.12 can be expressed through Equation 3.1.13 as follows:

$$\frac{d}{dt}\frac{\partial T}{\partial \dot{s}} - \frac{\partial T}{\partial s} - F^T Q^q = F^T \Gamma$$

$$\frac{d}{dt}\frac{\partial T}{\partial \dot{x}} - \frac{\partial T}{\partial x} - \Phi^T Q^q = \Phi^T \Gamma \qquad (3.1.14)$$

Taking into account Equations 3.1.5 and 3.1.14, as well as Equations 3.1.6 and 3.1.9, we transform Equations 3.1.12 to the following system:

$$F^T \Gamma = Q' + Q^* - \frac{1}{2}\epsilon^2 \frac{\partial (K(s)X, X)}{\partial s} \qquad (3.1.15)$$

$$\Phi^T \Gamma = Q^x - K(s)X - \epsilon^2 B(s)\dot{X} \qquad (3.1.16)$$

According to Equations 3.1.1 and 3.1.2 we have

$$\dot{q} = F\dot{s} + \Phi\dot{x} \tag{3.1.17}$$

Inserting the expressions given by Equation 3.1.3 for T, Equation 3.1.17 for $\dot{q}$, and Equation 3.1.9 for x into Equation 3.1.13, we obtain

$$\Gamma = \frac{d}{dt}[A(F\dot{s} + \epsilon^2\Phi\dot{X})] - \frac{1}{2}\frac{\partial(A(q)\dot{q}, q)}{\partial q} - Q^q \tag{3.1.18}$$

Hence, the motion of our mechanical system is described by Equations 3.1.15 and 3.1.16 and Equations 3.1.1, 3.1.2, 3.1.5, 3.1.9, 3.1.17, and 3.1.18.

Equations 3.1.15 and 3.1.16 with Γ given by Equation 3.1.18 comprise $n + m$ equations containing the second time derivatives of s and x.

Since the rank of matrix A is $N \leq n + m$, the order of simultaneous Equations 3.1.15 and 3.1.16 is equal to $2N$.

Let us discuss the assumptions in Equations 3.1.6 through 3.1.8. The assumptions of high stiffness (Equation 3.1.6) means that the periods T_i of natural elastic oscillations are of the order of ϵ: $T_i \sim \epsilon T_0$ where T_0 is the characteristic time of the rigid model motion. In the case of Equation 3.1.7 the elastic oscillations attenuate slightly: over the period T_0 of one oscillation the amplitude decreases by an amount of $O(\epsilon)$. In the case of Equation 3.1.8 the oscillations attenuate considerably: the amplitude decreases by $O(1)$ over the period T_0, and an aperiodic attenuation is possible.

The difference in behavior of systems obeying Equations 3.1.7 and 3.1.8 can be illustrated by the following simple example of the system with one degree of freedom. The equations

$$a\ddot{x} + \epsilon^{-2}kx + b\dot{x} = 0 \tag{3.1.19}$$

$$a\ddot{x} + \epsilon^{-2}kx + \epsilon^{-1}d\dot{x} = 0 \tag{3.1.20}$$

correspond to the assumptions expressed by Equations 3.1.6, 3.1.7 and Equations 3.1.6, 3.1.8, respectively. Here, a, k, b, and d are positive constants. The roots of the characteristic equations for Equations 3.1.19 and 3.1.20 are, respectively,

$$\lambda = \pm i\epsilon(k/a)^{1/2} - b(2a)^{-1} + O(\epsilon)$$

$$\lambda = (2\epsilon a)^{-1}[-d \pm 2(d^2 - 4ak)^{1/2}]$$

Therefore, for the system described by Equation 3.1.19 we have slightly damped oscillations (the real part of λ is negative and small compared to the imaginary part), while for the Equation 3.1.20 we have either strongly damped oscillations (if $d^2 < 4ak$) or an aperiodic motion (if $d^2 \geq 4ak$).

3.1.3. STATEMENT OF THE PROBLEMS

For our mechanical system described by Equations 3.1.15 and 3.1.16, relationships 3.1.1, 3.1.2, 3.1.5, 3.1.9, 3.1.17, and 3.1.18, and Conditions 3.1.7 or 3.1.8, we can consider the following problems of dynamics.

Problem 3.1. Motion in the case of kinematic control.

The law of change of the "rigid" generalized coordinates $s = s(t)$ is given. We are to find the motion of the system, i.e., the function $x(t)$, $q(t)$, and also the control forces $Q^*(t)$ that ensure the realization of $s(t)$.

Problem 3.2. Motion in the case of dynamic control.

The law of change of the control forces $Q^* = Q^*(t)$ is given. We are to find the motion of the system, i.e., the functions $s(t)$, $x(t)$, and $q(t)$.

Problem 3.3. Motion in the case of the prescribed change of "inertial" coordinates.

The law of change of the "inertial" coordinates $q = q(t)$ is given. We are to find the motion of the system, i.e., the functions $s(t)$, $x(t)$ and control forces $Q^*(t)$ that ensure the time history $q(t)$.

The initial data required for these problems will be stipulated separately for each case.

Note that Problem 3.2 belongs to the type of Problem 2.1, while Problems 3.1 and 3.3 belong to the type of Problem 2.2; see Chapter 2.

Problems 3.1 and 3.2 correspond to two different (and in some sense opposite) cases of control. If actuators of a robot can follow the prescribed kinematic law independently of the load we can consider Problem 3.1. In the case of Problem 2, the actuators must follow the prescribed law for torques and/or forces.

3.1.4. EXAMPLE

We consider the simplest electromechanical manipulator described in Chapter 2 (see Section 2.1.2 and Figure 2.1.1). If all elements of the robot are rigid, its motion is described by Equations 2.1.1, 2.1.10, and 2.1.13, i.e.,

$$I\ddot{\varphi} = M, \qquad L\dot{\mu} + R\mu - k_1 k_2 n\dot{\varphi} = k_2 u$$

$$M = -n\mu - I_0 n^2 \ddot{\varphi} \tag{3.1.21}$$

Here, all notations are specified in Section 2.1.2. Eliminating M and μ from Equations 3.1.21, we obtain the differential equation of the third order describing the motion of the manipulator arm:

$$LR^{-1}\dddot{\varphi} + \ddot{\varphi} + k_1 k_2 n^2 (I_* R)^{-1}\dot{\varphi} = -k_2 n (I_* R)^{-1} u$$

$$I_* = I + I_0 n^2 \tag{3.1.22}$$

Equation 3.1.22 contains two characteristic parameters having dimension of time:

$$\tau_1 = LR^{-1}, \qquad \tau_2 = I_* R/(k_1 k_2 n^2) \tag{3.1.23}$$

We introduce the dimensionless time $t' = t/T_0$, where T_0 is the characteristic time of the working operation. Then Equation 3.1.22 takes the form

$$\frac{\tau_1}{T_0}\frac{d^3\varphi}{dt'^3} + \frac{d^2\varphi}{dt'^2} + \frac{T_0}{\tau_2}\frac{d\varphi}{dt'} = -\frac{k_2 n T_0^2}{I_* R} u \tag{3.1.24}$$

where Equations 3.1.23 are used. If $\tau_1 \ll T_0$, then we can omit the first term in Equation 3.1.24, ignoring transient processes over the small time interval τ_1. Thus we obtain

$$\frac{d^2\varphi}{dt'^2} + \frac{T_0}{\tau_2}\frac{d\varphi}{dt'} = -\frac{k_2 n T_0^2}{I_* R} u \tag{3.1.25}$$

The parameter τ_1 is a characteristic damping time for processes in the electric circuit of the armature. Armatures of modern electric motors have usually low inductance, and the inequality $\tau_1 \ll T_0$ is thus fulfilled. The parameter τ_2, as is clear from Equation 3.1.25, is a transient period for the rotation of the manipulator arm. If $T_0 \ll \tau_2$, then Equation 3.1.25 can be simplified:

$$\ddot{\varphi} = -k_2 n (I_* R)^{-1} u$$

Here, we return to the dimensional time t. Comparing the obtained equation with the first Equation 3.1.21, we can see that the motion of our robot is governed by the equation

$$I\ddot{\varphi} = M, \qquad M = -I k_2 n (I_* R)^{-1} u$$

This case corresponds to the dynamic control of the manipulator.

If, on the contrary, $\tau_2 \ll T_0$, then we get from Equation 3.1.25

$$\dot{\varphi} = -(k_1 n)^{-1} u$$

In this case we can prescribe the angular velocity as well as the angle φ as functions of time by choosing the corresponding control $u(t)$. This case corresponds to the kinematic control.

3.1.5. RIGID MODEL

Let us consider the solution of Problems 3.1 to 3.3 in the limiting case of the rigid model ($\epsilon = 0$, $x = 0$). For kinematic control ($s = s(t)$) we obtain from Equations 3.1.1 and 3.1.15

$$x(t) \equiv 0, \qquad q(t) = f(s(t), 0) \equiv q^0(t)$$

$$Q^*(t) = (F^T \Gamma - Q')^0 \tag{3.1.26}$$

Here and henceforth in this chapter, the sign ^{0} means that the corresponding quantities are taken for the rigid model, i.e., for $x \equiv 0$, $s = s(t)$, $q = q^0(t)$. Equations 3.1.26 define the solution of Problem 3.1.

From Equation 3.1.16 we can find reaction forces R^x of those constraints that correspond to elastic elements as $\epsilon \to 0$. These forces are limits of the elastic forces as $\epsilon \to 0$. We obtain from Equations 3.1.16, 3.1.6, and 3.1.9,

$$R^x = -Cx = -KX = (\Phi^T \Gamma - Q^x)^0 \tag{3.1.27}$$

The work $(R^x, \delta x)$ of the forces given by Equation 3.1.27 on the virtual displacements δx is equal to zero, since $\delta x = 0$ for $\epsilon = 0$; the additional constraints imposed on the system as $\epsilon \to 0$ are ideal. From Equations 3.1.27 we obtain $X(t) = y(t)$, where

$$y(t) = [K^{-1}(Q^x - \Phi^T \Gamma)]^0 \tag{3.1.28}$$

Equation 3.1.28 defines quasistatic normalized (i.e., divided by ϵ^2, see Equation 3.1.9) elastic displacements which correspond to the reaction forces in Equation 3.1.27 and are bounded as $\epsilon \to 0$.

In the case of dynamic control (Problem 3.2), let us consider Equations 3.1.15 and 3.1.18 for $x = 0$, $\epsilon = 0$:

$$F^T \left[\frac{d}{dt}(AF\dot{s}) - \frac{1}{2} \frac{\partial(A(q)\dot{q}, \dot{q})}{\partial q} - Q^q \right] = Q' + Q^*$$

$$(x = 0, \qquad q = f(s(t), 0) \tag{3.1.29}$$

The rank of the matrix A is N, while the rank of the matrix F is $r = \min(n, N)$. Therefore, the order of the system of differential Equations 3.1.29 with respect to s is $2r$.

We consider the case of $N \geq n$ and $N < n$ separately. If $N \geq n$, then $r = n$, and the solution of Problem 3.2 reduce to integration of Equation 3.1.29 of order $2n$ for $s(t)$; the following initial conditions should be specified at the initial time instant $t = t_0$:

$$s(t_0) = s^1, \qquad \dot{s}(t_0) = s^2 \tag{3.1.30}$$

Here, s^1 and s^2 are given n-vectors. Having found $s(t)$, we can determine $q = f(s(t), 0)$, and this completes the solution of Problem 3.2.

If $N < n$, Equation 3.1.29 degenerates: its order is equal to $2N$. In this case, taking into account Equations 3.1.13 and 3.1.15, we write down the following system equivalent to Equation 3.1.29.

$$F^T \Gamma = Q' + Q^*(t), \qquad q = f(s, 0)$$

$$\frac{d}{dt} \frac{\partial T}{\partial \dot{q}} - \frac{\partial T}{\partial q} - Q^q = \Gamma \qquad\qquad (3.1.31)$$

This system of $2N + n$ equations for $2N + n$ unknowns s, q, Γ may have no solutions. Let, for the sake of simplicity, the forces Q' be independent of s and $\dot{s}$ and the matrix $F(s, 0)$ be constant. Then the first Equation 3.1.31 is a system of n linear algebraic equations for the N-dimensional vector Γ. This system is solvable only if some additional restrictions are imposed on the given control forces $Q^*(t)$. If these restrictions are satisfied, we can obtain the vector $\Gamma = \Gamma(t)$ from the first Equation 3.1.31. Then the third Equation 3.1.31 is a system of differential equations of order $2N$ for $q(t)$. Its solution $q(t)$ can be found if the initial conditions,

$$q(t_0) = q^1, \qquad \dot{q}(t_0) = q^2$$

are specified where q^1 and q^2 are given N-vectors. Then the vector s should be found from the second Equation 3.1.31 which contains N equations for $n > N$ unknowns. Hence, the vector $s(t)$ cannot be determined uniquely. Therefore, Problem 3.2 for $N < n$ for the rigid model can have an infinite number of solutions or no solutions at all. Henceforth we confine ourselves to the case $N \geqslant n$ in solving Problem 3.2.

In Problem 3.3, $q = q(t)$ is given, and we can find $\Gamma = \Gamma(t)$ from the third Equation 3.1.31. The control Q^* can be determined uniquely from the first Equation 3.1.31 if $s = s(t)$ is found. Thus, the solvability of Problem 3.3 is reduced to the possibility of finding $s(t)$ from the second Equation 3.1.31 for given q(t). If $N < n$, then $s(t)$ is, generally speaking, not unique; and if $N > n$, the solution for $s(t)$ exists only under certain additional conditions. Henceforth we restrict ourselves with the case of $N = n$ for Problem 3.3.

Denote by $s = h(q)$ the function inverse to $q = f(s, 0)$ for $N = n$. Then the solution of Problem 3.3 is defined by

$$s = h(q(t)), \qquad Q^* = F^T(s, 0)\Gamma(t) - Q'(s, \dot{s}, 0, 0, t) \quad (3.1.32)$$

Here, $\Gamma(t)$ is obtained from Equation 3.1.13 with $q = q(t)$.

Thus, in the case of the rigid model the solution of Problems 3.1 and 3.3 (for $N = n$) requires only algebraic operations and differentiation. The solution of Problem 3.2 (for $N \geqslant n$) reduces to integration of the system of differential Equations 3.1.29 under initial conditions specified by Equations 3.1.30.

In the next Section 3.2 we present solutions of Problems 3.1 to 3.3 for the initial system with elastic and dissipative elements ($\epsilon > 0$) in the case of small dissipation (Equation 3.1.7). We presume that solutions of the corresponding problems for the rigid model are known.

3.2. ANALYSIS OF THE CASE OF SMALL DISSIPATION

3.2.1. SOLUTION OF PROBLEM 3.1

We assume that the Condition 3.1.7 is satisfied. For specified $s = s(t)$ and taking into account Equations 3.1.1, 3.1.2, 3.1.5, 3.1.9, and 3.1.18, Equation 3.1.16 is a system of m equations for the m-dimensional vector function $X(t)$. The second derivative $\ddot{X}$ in these equations is multiplied by a small parameter, ϵ^2. In Equations 3.1.16 and 3.1.18 we make the change of variables:

$$X = y(t) + z(\tau), \qquad \tau = t/\epsilon \qquad (3.2.1)$$

Here, $y(t)$ is defined by Equation 3.1.28, and τ is a fast time. We substitute Equation 3.2.1 into Equations 3.1.16 and expand both sides of these equations in series in ϵ, using also Equations 3.1.2, 3.1.18 and $q = f(s, \epsilon^2 X)$. After a number of long, but straightforward manipulations we obtain

$$\frac{d}{d\tau}\left[M(t)\frac{dz}{d\tau}\right] + K^0(t)z = \epsilon G(t)\frac{dz}{d\tau}, \qquad t = \epsilon\tau \qquad (3.2.2)$$

Here, the terms $O(\epsilon^2)$ are omitted, and the following denotations are introduced:

$$M(t) = (\Phi^T A \Phi)^0, \qquad G(t) = G'(t) + G''(t)$$

$$G'_{ij}(t) = \sum_{k,l=1}^{N}\left[A^0_{kl}\left(\frac{d\Phi^0_{ki}}{dt}\Phi^0_{lj} - \frac{d\Phi^0_{lj}}{dt}\Phi^0_{ki}\right)\right.$$

$$+ \sum_{p=1}^{N}\frac{\partial A^0_{kl}}{\partial q_p}\frac{dq^0_k}{dt}\left.(\Phi^0_{pi}\Phi^0_{lj} - \Phi^0_{li}\Phi^0_{pj})\right]$$

$$G''_{ij}(t) = \sum_{k,l=1}^{N}\left(\frac{\partial Q^q_k}{\partial \dot{q}_l}\right)^0 \Phi^0_{kl}\Phi^0_{lj} + \left(\frac{\partial Q^s_i}{\partial \dot{x}_j}\right) - B^0_{ij}$$

$$q^0(t) = f(s(t), 0), \qquad K^0(t) = K(s(t)), \qquad i, j = 1, ..., m \qquad (3.2.3)$$

We will construct the analytic solution of the linear homogeneous Equation 3.2.2 in accordance with the asymptotic procedure described in References 86 and 169. First we consider, for fixed t, the eigenvalue problem,

$$(\lambda K^0 - M)\varphi = 0 \qquad (3.2.4)$$

Here, λ is an eigenvalue, φ is an m-dimensional eigenvector.

Since matrix A is positive definite while Φ has rank $\rho = \min(m, N)$, it follows from Equation 3.2.3 that the $m \times m$ matrix M is positive definite for $N \geqslant m$ and nonnegative definite for $N < m$. Matrix K^0 is positive definite (see Equation 3.1.6). Therefore, the characteristic equation corresponding to the problem in Equation 3.2.4,

$$\det(\lambda K^0 - M) = 0 \qquad (3.2.5)$$

has m real nonnegative roots. We assume for the sake of simplicity that all positive roots are simple. For $N \geqslant m$ all roots are positive, while for $N < m$ there are N positive roots and a zero root of multiplicity $m - N$. In all cases there are m linearly independent eigenvectors φ^k, $k = 1, \ldots, m$. Let $\lambda_1, \ldots, \lambda_\rho$ be positive roots of characteristic Equation 3.2.5, while $\varphi^1, \ldots, \varphi^\rho$ are the corresponding eigenvectors satisfying Equations 3.2.4. Then the asymptotic solution of the first approximation of Equation 3.2.2 can be presented as an expansion in normal oscillations:

$$z(\tau) = \sum_{k=1}^{\rho} \varphi^k a_k \cos \psi_k, \qquad \rho = \min(m, N) \qquad (3.2.6)$$

The frequencies and phases of the normal oscillations are given by the expressions

$$\omega_k(t) = \lambda_k^{-1/2}(t)$$

$$\psi_k(t) = \frac{1}{\epsilon} \int_{t_0}^{t} \omega_k(t_1)dt_1 + \psi_{k0}, \qquad k = 1, \ldots, \rho \qquad (3.2.7)$$

while the amplitudes satisfy the differential equations

$$\frac{da_k}{dt} = -\frac{\epsilon a_k}{2\mu_k \omega_k} \frac{d(\mu_k \omega_k)}{dt}$$

$$+ \frac{\epsilon}{2\pi\mu_k \omega_k} \int_0^{2\pi} (G\varphi^k \omega_k a_k \sin \psi, y^k) \sin \psi \, d\psi$$

$$= -\frac{\epsilon a_k}{2} \frac{d \ln(\mu_k \omega_k)}{dt} + \frac{\epsilon \nu_k a_k}{2\mu_k}, \qquad k = 1, \ldots, \rho \qquad (3.2.8)$$

Here, the following denotations are introduced:

$$\mu_k(t) = (\varphi^k)^T M \varphi^k$$

$$\nu_k(t) = (\varphi^k)^T G \varphi^k = (\varphi^k)^T G'' \varphi^k, \qquad k = 1, \ldots, p \qquad (3.2.9)$$

The last equality in Equations 3.2.9 stems from the fact that matrix G' in Equation 3.2.3 is skew-symmetric. Equations 3.2.8 are integrated in quadratures:

$$a_k(t) = a_{k0} \left[\frac{\mu_k(t_0)\omega_k(t_0)}{\mu_k(t)\omega_k(t)} \right]^{1/2} \exp \int_{t_0}^{t} \frac{\nu_k(t_1)dt_1}{2\mu_k(t_1)}$$

$$k = 1, \ldots, p \qquad (3.2.10)$$

To calculate the solution given by Equation 3.2.6 it is necessary to solve the eigenvalue problem, i.e., to find λ_k from Equation 3.2.5 and φ^k from Equation 3.2.4, and then to determine ω_k, ψ_k, and a_k using Equations 3.2.7, 3.2.9, and 3.2.10. Since matrices K^0 and M depend on t, it follows that λ_k, ω_k, and φ^k also depend on t. If the forces Q^q and Q^x are independent of the velocities and the Rayleigh function R is equal to zero ($\partial Q^q/\partial \dot{q} = \partial Q^x/\partial \dot{x} = B = 0$), we have $G'' = 0$ (see Equation 3.2.3). The Equation 3.2.10 is simplified in this case of adiabatic invariance:

$$a_k(t) = a_{k0} \left[\frac{\mu_k(t_0)\omega_k(t_0)}{\mu_k(t)\omega_k(t)} \right]^{1/2}, \qquad k = 1, \ldots, p$$

Asymptotic solution given by Equation 3.2.6 defines z with an error $O(\epsilon)$ on the time interval $\Delta t = t - t_0 = O(1)$. For the derivative we obtain from Equations 3.2.6 and 3.2.7,

$$\dot{z} = -\epsilon^{-1} \sum_{k=1}^{p} \varphi^k a_k \omega_k \sin \psi_k + O(1) \qquad (3.2.11)$$

In accordance with Equations 3.1.9, 3.2.1, 3.2.6, and 3.2.11 the total elastic displacement x and its time derivative are

$$x = \epsilon^2(y + z) + O(\epsilon^3)$$

$$\dot{x} = -\epsilon \sum_{k=1}^{p} \varphi^k a_k \omega_k \sin \psi_k + O(\epsilon^2) \qquad (3.2.12)$$

According to Equations 3.1.1 and 3.2.12, vector q is

$$q = f(s(t), x) = q^0(t) + \epsilon^2 \Phi^0(t)(y + z) + O(\epsilon^3) \qquad (3.2.13)$$

The solution obtained (Equation 3.2.6) depends on the $2p$ arbitrary constants a_{ko} and ψ_{ko} that appear in Equations 3.2.7 and 3.2.10. Their number is equal to the order of Equation 3.2.2, since the rank of matrix M is equal to p. To determine these constants, we are to specify the initial elastic displacements $x(t_0)$ and velocities $\dot{x}(t_0)$; in accordance with Equations 3.2.12, these should be of order $O(\epsilon^2)$ and $O(\epsilon)$, respectively. Then we should employ Equations 3.2.6 and 3.2.12, as a result of which we obtain a linear algebraic system of $2m$ equations for the $2p$ unknowns $a_{ko}\cos\psi_{ko}$ and $a_{ko}\sin\psi_{ko}$. In the case of $N \geq m$ we have $\rho = m$, and this system is solvable. For $N < m$, $\rho = N$, the initial data cannot be arbitrary: they should obey the $m - N$ linear constraints required for the system to be solvable.

To determine the control forces Q^*, we substitute the solutions given by Equations 3.2.12, 3.2.6, and 3.2.13 into Equation 3.1.15. After simplification, we obtain

$$Q^*(t) = (F^T\Gamma - Q')^0 - (F^T A \Phi)^0 \sum_{k=1}^{\rho} \varphi^k a_k \omega_k^2 \cos\psi_k + O(\epsilon) \quad (3.2.14)$$

Here, the first term is equal to forces in Equation 3.1.26 in the case of the rigid model, while the second corresponds to fast elastic oscillations.

Thus we have completed the solution of Problem 3.1. The elastic displacements and velocities (Equations 3.2.12), inertial coordinates (Equation 3.2.13), and forces (Equation 3.2.14) consist of slow (quasistatic) and fast (oscillatory) terms; the fast terms in the Equations 3.2.12 for displacements and Equation 3.2.14 for forces are of the same order as the slow terms. In Equations 3.2.12 for velocities, the fast terms are greater than the slow ones.

3.2.2. SOLUTION OF PROBLEM 3.2 FOR $N = n + m$

Problem 3.2 of dynamic control will be considered in two cases: for $N = n + m$ in Section 3.2.2 and for $N = n$ in Section 3.2.3. In the case $N = n + m$ for specified forces $Q^*(t)$, Equations 3.1.15 and 3.1.16 form a system of N equations for N unknowns s and X. We seek the solution of this system of order $2N$ in the form

$$s = s^0(t) + \epsilon^2 u(\tau), \qquad X = y^0(t) + z(\tau), \qquad \tau = t/\epsilon \quad (3.2.15)$$

Here, u and z are new unknown functions, while $s^0(t)$ is a solution of Equation 3.1.15 for the rigid model, i.e., of Equation 3.1.29. We mentioned in Section 3.1.5 that Equation 3.1.29 for $N \geq n$ is a system of order $2n$ for $s(t)$. Solving it with initial conditions (Equation 3.1.30) we can obtain $s^0(t)$. In Equation 3.2.15, $y^0(t)$ denotes the function given by Equation 3.1.28 in which $s = s^0(t)$. We substitute s and x from Equation 3.2.15 into Equations 3.1.15 and 3.1.16 and expand both sides of these equations in series in ϵ, using Equations 3.1.1, 3.1.2, 3.1.15, 3.1.9, 3.1.28, and 3.1.29. After cumbersome manipulations, and omitting terms $O(\epsilon^2)$, we obtain a linear homogeneous system for the functions u and z that is analogous to Equation 3.2.2:

$$\frac{d}{d\tau}\left[P(t)\frac{du}{d\tau} + S(t)\frac{dz}{d\tau}\right] = \epsilon H(t)\frac{du}{d\tau} + \epsilon J^1(t)\frac{dz}{d\tau}, \qquad t = \epsilon\tau$$

$$\frac{d}{d\tau}\left[S^T(t)\frac{du}{d\tau} + M(t)\frac{dz}{d\tau}\right] + K^0(t)z = \epsilon^2 J^2(t)\frac{du}{d\tau} + \epsilon G(t)\frac{dz}{d\tau} \qquad (3.2.16)$$

Here, we have used Equations 3.2.3 and have introduced the following quantities:

$$P(t) = (F^T A F)^0, \qquad S(t) = (F^T A \Phi)^0, \qquad H = H' + H''$$

$$J^1 = J + J', \qquad J^2 = -J^T + J''$$

$$H'_{ij}(t) = \sum_{k,l=1}^{N}\left[A^0_{kl}\left(\frac{dF^0_{ki}}{dt}F^0_{lj} - \frac{dF^0_{lj}}{dt}F^0_{ki}\right) + \sum_{p=1}^{N}\frac{\partial A^0_{kl}}{\partial q_p}\frac{dq^0_k}{dt}(F^0_{pi}F^0_{lj} - F^0_{li}F^0_{pj})\right]$$

$$H''_{ij}(t) = \sum_{k,l=1}^{N}\left(\frac{\partial Q^q_k}{\partial \dot q_l}\right)F^0_{ki}F^0_{lj} + \left(\frac{\partial Q'_i}{\partial \dot s_j}\right)^0$$

$$J_{ij}(t) = \sum_{k,l=1}^{N}\left[A^0_{kl}\left(\frac{dF^0_{ki}}{dt}\Phi^0_{lj} - \frac{d\Phi^0_{lj}}{dt}F^0_{ki}\right) + \sum_{p=1}^{N}\frac{\partial A^0_{kl}}{\partial q_p}\frac{dq^0_k}{dt}(F^0_{pi}\Phi^0_{lj} - F^0_{li}\Phi^0_{pj})\right]$$

$$J'_{ij}(t) = \sum_{k,l=1}^{N}\left(\frac{\partial Q^q_k}{\partial \dot q_l}\right)F^0_{ki}\Phi^0_{lj} + \left(\frac{\partial Q'_i}{\partial \dot x_j}\right)^0$$

$$J''_{ij}(t) = \sum_{k,l=1}^{N}\left(\frac{\partial Q^q_k}{\partial \dot q_l}\right)F^0_{lj}\Phi^0_{ki} + \left(\frac{\partial Q^x_i}{\partial \dot s_j}\right)^0$$

$$s = s^0(t), \qquad x = 0, \qquad q = q^0(t) = f(s^0(t), 0) \qquad (3.2.17)$$

Matrices P and M in Equations 3.2.16 and 3.2.17 are positive definite. Using the symmetry of matrix A, we readily establish that H' is a skew-symmetric matrix.

Let us now proceed to construct the asymptotic solution of Equations 3.2.16. We integrate both sides of the first Equation 3.2.16 with respect to τ, integrating its right side by parts (c is an arbitrary constant):

$$P(t)\frac{du}{d\tau} + S(t)\frac{dz}{d\tau} = \epsilon[H(t)u + J^1(t)z]$$

$$+ \epsilon^2\int_{\tau_0}^{\tau}\left[\frac{dH(t_1)}{dt_1}u(t_1) + \frac{dJ^1(t_1)}{dt_1}z(t_1)\right]d\tau_1 + c$$

$$\tau = t/\epsilon, \qquad \tau_1 = t_1/\epsilon, \qquad \tau_0 = t_0/\epsilon$$

It follows from the obtained equation that for the interval $t - t_0 = O(1)$, $\tau - \tau_0 = O(\epsilon^{-1})$ we have

$$P(t) \frac{du}{d\tau} + S(t) \frac{dz}{d\tau} = c + O(\epsilon) \tag{3.2.18}$$

If $c = O(1)$, then, integrating Equation 3.2.18 we obtain $Pu + Sz = O(\epsilon^{-1})$ for the interval $\tau - \tau_0 = O(\epsilon^{-1})$, and this contradicts the boundedness of u and z. Consequently, we set $c = c_1\epsilon$ (c_1 is a new constant); then Equation 3.2.18 yields

$$\frac{du}{d\tau} = -P^{-1}S \frac{dz}{d\tau} + O(\epsilon) \tag{3.2.19}$$

Now we solve the first equation in Equations 3.2.16 with respect to d^2u/dt^2, and in the resultant equation we express du/dt in accordance with Equation 3.2.19:

$$\frac{d^2u}{d\tau^2} = -P^{-1} \frac{d}{d\tau}\left(S \frac{dz}{d\tau} \right)$$
$$+ \epsilon P^{-1}(\dot{P}P^{-1}S - HP^{-1}S + J^1) \frac{dz}{d\tau} + O(\epsilon^2) \tag{3.2.20}$$

We differentiate in the left side of the second equation in Equations 3.2.16 and insert Equations 3.2.19 and 3.2.20 for the derivatives of u into it. After some manipulations we obtain, within the accuracy of $O(\epsilon^2)$,

$$\frac{d}{dt}\left[M^*(t) \frac{dz}{d\tau} \right] + K^0(t)z = \epsilon G^*(t) \frac{dz}{d\tau}$$
$$M^*(t) = M - S^T P^{-1} S$$
$$G^*(t) = S^T P^{-1} H P^{-1} S - S^T P^{-1} J^1 - J^2 P^{-1} S + G \tag{3.2.21}$$

We will prove that the symmetric matrix M^* introduced in Equation 3.2.21 is positive definite. For this we consider the quadratic form of $n + m$ variables:

$$\xi^T P \xi + \eta^T M \eta + \xi^T S \eta + \eta^T S^T \xi \tag{3.2.22}$$

where ξ and η are n-dimensional and m-dimensional vectors, respectively. Taking into account Equations 3.2.3 and 3.2.17 for matrices M, P, and S, we see that the form in Equation 3.2.22 corresponds to kinetic energy (Equation 3.1.3) of our system. Hence, it is positive definite in the case $N = n + m$ under consideration. Substituting $\xi = -P^{-1}S\eta$ in Equation 3.2.22, we obtain the positive definite quadratic form $\eta^T M^* \eta$, and this proves that M^* is a positive definite matrix.

The asymptotic solution of Equation 3.2.21 is constructed exactly in the same way as in Section 3.2.1, and, similarly to Equation 3.2.6, it has the form (here $\rho = m$):

$$z(\tau) = \sum_{k=1}^{m} \varphi^k a_k \cos \psi_k \tag{3.2.23}$$

Here, ψ_k, a_k are defined by Equations 3.2.7 and 3.2.10 with $\rho = m$. The quantities μ_k and ν_k in these formulas are given by expressions similar to Equations 3.2.9:

$$\mu_k(t) = (\varphi^k)^{\mathrm{T}} M^* \varphi^k, \qquad \nu_k(t) = (\varphi^k)^{\mathrm{T}} G^* \varphi^k$$

$$= (\varphi^k)^{\mathrm{T}} [S^T P^{-1} H'' P^{-1} S - S^T P^{-1} J' - J'' P^{-1} S + G''] \varphi^k$$

$$k = 1, \dots, m \tag{3.2.24}$$

Here, we have employed Equations 3.2.21, 3.2.17, and 3.2.3, in which H' and G' are skew-symmetric matrices. The frequencies ω_k and eigenvectors φ^k are determined from the eigenvalue problem similar to Equations 3.2.4 and 3.2.5:

$$\det(K^0 - \omega_k^2 M^*) = 0, \qquad (K^0 - \omega_k^2 M^*) \varphi^k = 0$$

$$K^0 = K(S^0(t)), \qquad k = 1, \dots, m \tag{3.2.25}$$

Since the matrix M^* is positive definite, characteristic Equation 3.2.25 always had m positive roots ω_k, which we assume to be simple.

Vectors x and $\dot{x}$ are given by the earlier formulas (Equations 3.2.12) in which $y = y^0(t)$, $z(t)$ is defined by Equation 3.2.23, and $\rho = m$. The arbitrary constants a_{k0} and ψ_{k0}, in the solutions given by Equations 3.2.23, 3.2.7, and 3.2.10, can be determined as in Section 3.2.1. As $\rho = m$, the problem of determining these constants is solvable.

The function $u(\tau)$ appears in the Equation 3.2.15 for s. It follows from Equation 3.2.19 that $du/d\tau$ is determined with an error $O(\epsilon)$ in the approximation under consideration, and therefore u is determined with error $O(1)$ on the interval $\tau - \tau_0 = O(\epsilon^{-1})$. Consequently we have, in accordance with Equations 3.2.15 and 3.1.1,

$$s = s^0(t) + O(\epsilon^2), \quad q = q^0(t) + O(\epsilon^2), \quad q^0 = f(s^0(t), 0) \tag{3.2.26}$$

Here, $s^0(t)$ is the solution of Equation 3.1.29 with initial conditions given by Equations 3.1.30. Thus, in the case $N = n + m$ under consideration, the solution of Problem 3.2 in the first approximation is given by Equations 3.2.12 and 3.2.26 in which y and z are defined by Equations 3.1.28 and

3.2.23, respectively. In contrast with Section 3.2.1 (see Equation 3.2.13), in this case the vector q is defined with a lower degree of accuracy than x.

3.2.3. SOLUTION OF PROBLEM 3.2 FOR $N = n$

In contrast with the general case (see Equations 3.1.5), in this section we assume that the forces Q' and Q^x do not depend on velocities, so that

$$Q' = Q'(s, x, t), \qquad Q^x = Q^x(s, x, t)$$
$$Q^s = Q' + Q^* = Q^s(s, x, t) \tag{3.2.27}$$

and, in addition, that $B = 0$. For $N = n$, F is an $n \times n$-matrix, and its rank is n. Therefore, Equation 3.1.15 can be solved for vector Γ. Taking account of Equations 3.2.27, we obtain

$$\Gamma = (F^{-1})^T \left[Q^s - \frac{1}{2} \epsilon^2 \frac{\partial (X^T K X)}{\partial s} \right] \tag{3.2.28}$$

We substitute Γ from Equation 3.2.28 into Equation 3.1.16 and solve the resultant equation for X:

$$X = K^{-1}(s)\{Q^x(s, 0, t) - \Phi(s, 0)[F^{-1}(s, 0)]^T Q^s(s, 0, t)\} \tag{3.2.29}$$

Here, we have omitted terms $O(\epsilon^2)$. In the case of the rigid model we have, instead of Equation 3.2.28,

$$\Gamma^0 = [(F^0)^{-1}]^T Q^s(s^0, 0, t), \qquad q^0 = f(s^0(t), 0) \tag{3.2.30}$$

For $\epsilon = 0$, $x = 0$, Equations 3.2.30, together with Equation 3.1.18, form a system of n equations of order $2n$ for the vector-valued function $s^0(t)$. Solving it with initial conditions given by Equation 3.1.30 we can determine $s^0(t)$.

To solve the initial system, we can proceed in the following natural fashion. Substituting X from Equation 3.2.29 and Γ from Equation 3.1.18 into Equation 3.2.28, we obtain a system of n equations for the function $s(t)$, whose solution can be sought in the form $s(t) = s^0(t) + \epsilon^2 \alpha(t)$. For $\alpha(t)$ we obtain a linear system of order $2n$. Thus, to determine $s(t)$ with an error $O(\epsilon)$, it is necessary to integrate two systems of order $2n$ (for s^0 and α).

Let us present another approach to the solution, one that requires integration of only one system of order $2n$. We introduce an auxiliary rigid model that is equivalent to the initial system in the sense that identically with respect to t we have

$$f(s(t), x(t)) = f(\sigma(t), 0) \tag{3.2.31}$$

Here, σ is the vector of generalized coordinates of the auxiliary model. In Equation 3.2.31 we set

$$s = \sigma + \epsilon^2\alpha + O(\epsilon^4), \qquad x = \epsilon^2 X \qquad (3.2.32)$$

Now we expand the left-hand side of Equation 3.2.31 in a series in ϵ^2:

$$f(\sigma, 0) + \epsilon^2[F(\sigma, 0)\alpha + \Phi(\sigma, 0)X] + O(\epsilon^4) = f(\sigma, 0)$$

From this we obtain, taking account of Equation 3.2.29,

$$\alpha = -F^{-1}(\sigma, 0)\Phi(\sigma, 0)X(\sigma, t)$$

$$X(\sigma, t) = K^{-1}(\sigma)\{Q^s(\sigma, 0, t) - \Phi^T(\sigma, 0)[F^{-1}(\sigma, 0)]^T Q^s(\sigma, 0, t)\} \qquad (3.2.33)$$

Here, vector s is replaced by σ in the expression for X; this does not impair the accuracy of determining s assumed in Equation 3.2.32. Equations 3.2.32 and 3.2.33 establish the relationship between s and σ. Note that Equation 3.2.31 ensures that the coordinates $q(t)$ of the initial and auxiliary systems coincide, and hence so do the vectors Γ for them, since Γ, according to Equation 3.1.13, depends only on $q(t)$. We determine vector Γ^σ for the auxiliary system on the basis of Equation 3.2.30 for the rigid model, in which we replace s^0 by σ, and Q^s by the yet unknown force $Q^\sigma(\sigma, t)$. Equating Γ^σ to vector Γ from Equation 3.2.28, we obtain

$$[F^{-1}(s, \epsilon^2 X)]^T\left[Q^s(s, \epsilon^2 X, t) - \frac{1}{2}\epsilon^2\frac{\partial(X^T K(s)X)}{\partial s}\right]$$

$$= [F^{-1}(\sigma, 0)]^T Q^\sigma(\sigma, t)$$

We substitute s from Equation 3.2.32 into this equation, expand its left-hand side in a series in ϵ^2, and determine Q^σ:

$$Q^\sigma(\sigma, t) = Q^s(\sigma, 0, t) + \epsilon^2 F^T(\sigma, 0)\left\{\frac{\partial[(F^{-1}(\sigma, 0))^T Q^s(\sigma, 0, t)]}{\partial\sigma}\,\alpha\right.$$

$$+ \left.\frac{\partial[(F^{-1}(\sigma, x))^T Q^s(\sigma, x, t)]}{\partial x}\bigg|_{x=0}X\right\}$$

$$- \frac{1}{2}\epsilon^2\sum_{i,j=1}^m\frac{\partial K_{ij}(\sigma)}{\partial\sigma}X_iX_j + O(\epsilon^4) \qquad (3.2.34)$$

The force Q^σ for the auxiliary model differs from the force $Q^s(s, 0, t)$ for the rigid model by additional terms; it follows from Equations 3.2.34 and 3.2.33 that these terms depend quadratically on the forces Q^s and Q^x.

Consequently, the solution of Problem 3.2 can be represented as follows. We successively compute functions $X(\sigma, t)$, $\alpha(\sigma, t)$, and $Q^\sigma(\sigma, y)$ using

Equations 3.2.33 and 3.2.34. Then we set up the equations of motion of the auxiliary rigid model in the form of Equation 3.2.30, replacing the variable s^0 in them by σ and the force Q^s by Q^σ. Thus we obtain the system similar to Equation 3.1.29:

$$F^T(\sigma, 0)\left\{\frac{d}{dt}\,[A(q)F(\sigma, 0)\dot\sigma] - \frac{1}{2}\,\frac{\partial A((q)\dot q, \dot q)}{\partial q} - Q^q\right\} = Q^\sigma$$

$$q = f(\sigma, 0) \qquad\qquad\qquad\qquad (3.2.35)$$

The obtained Equation 3.2.35 of order $2n$ for $\sigma(t)$ should be integrated with the initial conditions,

$$\sigma(t_0) = s^1 - \epsilon^2\alpha(s^1, t_0)$$

$$\dot\sigma(t_0) = s^2 - \epsilon^2\left[\frac{\partial\alpha(s^1, t_0)}{\partial t} + \frac{\partial\alpha(s^1, t_0)}{\partial\sigma}\,s^2\right] \qquad (3.2.36)$$

which follow from Equations 3.1.30 and the relationship between s and σ given by Equation 3.2.32. Once we have determined $\sigma(t)$, we obtain successively X, α, s, and x as functions of t using Equations 3.2.33 and 3.2.32. Furthermore, we have, according to Equation 3.2.31,

$$q = f(s, x) = f(\sigma(t), 0) \qquad\qquad (3.2.37)$$

Thus, all the unknown functions $s(t)$, $x(t)$, and $q(t)$ in Problem 3.2 are determined with an error $O(\epsilon^4)$. We should point out that this case differs markedly from Sections 3.2.1 and 3.2.2. First, there are no high-frequency elastic oscillations here, and, second, all the unknowns are determined with a higher degree of accuracy. It is necessary here to integrate numerically the equations of motion of the auxiliary rigid model only once.

We have described the solution of Problem 3.2 in the case of $N = n + m$ and $N = n$. For $n < N < n + m$ the motion is similar to the case of $N = n + m$, but the solution becomes more complicated.

3.2.4. SOLUTION OF PROBLEM 3.3

In Problem 3.3, according to Section 3.1.3, we consider the function $q(t)$ as specified (known) and assume $n = N$ in accordance with Section 3.1.5. The initial values of coordinates $q(t_0)$ and velocities $\dot q(t_0)$ are assumed to correspond to the given function $q(t)$. For the sake of simplicity, we suppose that Equations 3.2.27 and the condition $B = 0$ imposed in Section 3.2.3 are also fulfilled here.

We seek the solution for s and x in the form of Equation 3.2.32. Substituting s from Equation 3.2.32 into the identity $q(t) = f(s, x)$ and expanding its right side in a series in ϵ^2, we obtain

$$\sigma(t) = h(q(t)), \qquad \alpha(t) = -F^{-1}(\sigma, 0)\Phi(\sigma, 0)X \qquad (3.2.38)$$

Here, $\sigma = h(q)$ is a function inverse to $q = f(\sigma, 0)$, which was introduced in Equation 3.1.32 for the rigid model. We obtain from Equation 3.1.16

$$X(t) = K^{-1}(\sigma)[Q^x(\sigma, 0, t) - \Phi^T(\sigma, 0)\Gamma(t)] + O(\epsilon^2) \qquad (3.2.39)$$

Here, $\Gamma(t)$ is determined by Equation 3.1.13, and Equation 3.2.32 as well as condition $B = 0$ and Equation 3.2.27 are taken into account.

Now we insert Equations 3.2.27 and 3.2.32 into Equation 3.1.15 and obtain the control forces Q^*:

$$
\begin{aligned}
Q^*(t) &= F^T(\sigma + \epsilon^2\alpha, \epsilon^2 X)\Gamma(t) - Q'(\sigma + \epsilon^2\alpha, \epsilon^2 X, t) \\
&\quad + \frac{1}{2}\epsilon^2 \frac{[X^T K(\sigma)X]}{\partial\sigma} \\
&= [F^T(\sigma, 0)\Gamma(t) - Q'(\sigma, 0)] + \epsilon^2 Q^{**} + O(\epsilon^4) \qquad (3.2.40)
\end{aligned}
$$

The components of the vector Q^{**} in Equation 3.2.40 are given by

$$
\begin{aligned}
Q_k^{**}(t) &= \sum_{i=1}^{n}\left[\sum_{j=1}^{n}\frac{\partial F_{ik}(\sigma, 0)}{\partial\sigma_j}\alpha_j + \sum_{j=1}^{n}\frac{\partial F_{ik}(\sigma, x)}{\partial x_j}\bigg|_{x=0}X_j\right]\Gamma_i(t) \\
&\quad - \sum_{j=0}^{n}\frac{\partial Q_k'(\sigma, 0, t)}{\partial\sigma_j}\alpha_j - \sum_{j=1}^{m}\frac{\partial Q_k'(\sigma, x, t)}{\partial x_j}\bigg|_{x=0}X_j \\
&\quad + \frac{1}{2}\frac{\partial[X^T K(\sigma)X]}{\partial\sigma_k}, \qquad k = 1, ..., n \qquad (3.2.41)
\end{aligned}
$$

Thus, the solution of Problem 3.3 is reduced to the following successive operations. For a given $q(t)$, we determine $\sigma(t)$ from Equation 3.2.38 and $\Gamma(t)$ from Equation 3.1.13. Then we find $X(t)$ from Equation 3.2.39, $\alpha(t)$ from Equation 3.2.38, and $Q^{**}(t)$ from Equation 3.2.41. Then the vectors s, x, and Q^* can be obtained by means of Equations 3.2.32 and 3.2.40 with an error $O(\epsilon^4)$. The procedure described gives the solution that differs by terms of order $O(\epsilon^2)$ from the solution of Problem 3.3 (Equation 3.1.32) obtained for the rigid model. To construct the solution, we should perform only algebraic operations and differentiation, as in the case of the rigid model. Note that the high-frequency elastic oscillations do not appear here, similar to Section 3.2.3.

3.2.5. CONCLUSION

We have described above the asymptomatic procedure developed in References 57 and 58 for the solution of Problems 3.1 to 3.3 in the case of small damping (Equation 3.1.7). The case of strong damping (Equation 3.1.8) was also investigated by means of asymptotic methods of small parameter in References 57 and 58 (see also References 59, 60, and 69). However, in the case of strong damping another type of asymptotic method was used, namely, methods of boundary layer, or singular perturbations.[171,238] These methods were used before for investigations in dynamics of rigid bodies containing cavities with viscous fluid[53,54] or carrying elastic and dissipative elements.[55,56,60,64]

The approach developed in this chapter may be called semianalytical, since part of the results are obtained in an explicit analytical form, while in some cases it is also necessary to integrate numerically certain systems of differential equations. However, it is important to stress that the equations that should be integrated numerically are equivalent to the equations of the rigid model. These equations do not include high-frequency elastic oscillations, since the fast motions are described by analytical formulas. Consequently, employing the approach developed above, we can choose comparatively large steps of numerical integration (the same as for the rigid model), while using the direct numerical approach requires much smaller integration steps. Thus, the asymptotic approach permits reducing significantly the time of numerical simulation. We used the Lagrange equations in this chapter; however, the asymptotic procedure developed above can be applied also to the other form of dynamical equations.

Chapter 4

DYNAMICS OF MANIPULATION ROBOTS WITH ELASTIC LINKS

4.1. ROBOTS WITH ELASTIC ELEMENTS (A SURVEY)

In Chapters 4 through 6 we deal with dynamics of manipulators with elastic parts (links and/or joints). The elastic compliance can essentially affect dynamic properties and performance characteristics of a robot. Due to elastic displacements, joint angles do not uniquely determine the position and orientation of the gripper. This causes errors in measuring parameters of the gripper motion by sensors placed at the joints and therefore reduces the accuracy of control. The elasticity of the structure of a robot can cause unwanted oscillations and resonant effects that reduce the accuracy and reliability of the system, complicate the control, and sometimes lead to damages in the structure. Therefore, the mechanical part of a robot and its information and control systems must be designed so as to mitigate undesirable consequences of the elastic compliance.

In the majority of industrial robots, however, the elastic compliance is not very significant except for extremely precise operations. This is achieved by making links of robots rather thick and massive to increase their bending and torsional stiffnesses. Thus, eigenfrequencies of the structure of a robot are far beyond the frequencies of the control loop and external disturbances. However, this requires a rather significant material expenditure, powerful drives, and excessive energy consumption. Today, the ratio of the payload mass to that of a robot amounts from 0.1 to 0.03 and less.[142] It is about 10 to 20 times as much as strength of materials requires.

An important scientific and engineering problem of modern robotics is to reduce the mass of robot structure and, at the same time, to ensure high accuracy and performance characteristics. The reduction of the mass will lead to a significant elastic compliance, and the mathematical models with rigid links will not suffice. High accuracy and productivity of such flexible manipulators can be achieved only if the control system is designed so as to suppress the elastic oscillations.

Though structures of modern manipulators are usually stiff, the elastic compliance is nevertheless significant in some applications, especially for estimating positioning accuracy. For industrial robots, flexibility is important in such operations as spot welding or assembly. Experiments show (see, e.g., Chapter 5 of this book) that the elastic compliance of industrial robots is mostly due to the compliance of joints and gear trains, the contribution of distributed elasticity of arm links being much less. On the other hand, the compliance of large robots (e.g., space manipulators) is caused mostly by the elasticity of the links.

The investigation in dynamics and control of manipulators with elastic parts is an important scientific direction in robotics. This is confirmed by a large number of publications on this issue. These publications cover the following basic fields of research: (1) the analysis of natural elastic oscillations, their frequencies, and shapes; (2) the mathematical simulation of the dynamics of elastic robots; and (3) the design of control laws, taking into account the elastic compliance.

The spectrum of natural elastic oscillations of a robot is an important property of the system. The knowledge of eigenfrequencies allows designing the mechanical and control systems of the robot so as to avoid resonant relationships. If, in addition, the shapes of natural oscillations are known, then we can calculate the motion of the elastic manipulator by using the modal approach.[44,142,162,175]

The natural oscillations of manipulators are investigated, for example, in References 143, 165, 168, 213. In Reference 213, an approximate approach for estimating eigenfrequencies of robots with elastic links is presented. The eigenfrequencies and oscillation shapes of a two-link manipulator with inertia-free links and a point payload are obtained in Reference 165 (see also Section 4.3 of this book). These results are extended in Reference 168 to a manipulator in which both the links and joints are elastic.

Numerous papers are devoted to mathematical modeling and simulation of robots with elastic parts. Surveys of this field, together with reviews of relevant publications, can be found in References 77, 78, 142. In the general case, when both the links and joints of a robot are elastic, a complete mathematical model of the system is a set of differential and/or integrodifferential equations containing ordinary and partial derivatives, with corresponding boundary conditions. However, the majority of authors reduce the model to a finite system of ordinary differential equations, thus considering a manipulator as a mechanical system with a finite number of degrees of freedom. The basic tools for such reduction are the finite element method and the modal approach.

When using the finite element method, a continuous elastic body (e.g., the link of a manipulator) is approximately treated as a system of discrete parts (finite elements). Each of these parts is described by a finite number of generalized coordinates. In the general case, elastic and dissipative forces act between the finite elements. Note that for the same mechanical system, finite elements can be chosen in different ways, and the success in using this approach depends on that choice. The finite element method is applied to modeling multilink elastic mechanisms, in particular manipulation robots, e.g., in References 93, 126, 148, 164, 221, 222.

The modal approach consists in representing the elastic displacement $u_i(x, t)$ of the ith link as a linear combination:

$$u_i(x, t) = \sum_{j=1}^{n_i} q_{ij}(t)S_{ij}(x) \qquad (4.1.1)$$

Here, $q_{ij}(t)$, $j = 1, \ldots, n_i$ are time-varying coefficients ("elastic" generalized coordinates of the ith link), and $S_{ij}(x)$, $j = 1, \ldots, n_i$ are vector-valued functions of spatial coordinates x of the ith link. The functions $S_{ij}(x)$ and the number of terms, n_i, in Equation 4.1.1 can be chosen in different ways. As $S_{ij}(x)$, we may take the eigenfunctions obtained from the mathematical model based on partial differential equations and boundary conditions. In this case $S_{ij}(x)$ determines the shapes of oscillation modes of the manipulator. The described approach is used, for example, in References 20, 200. However, explicit expressions for eigenfunctions can be found only for relatively simple structures. Therefore, other functions of $S_{ij}(x)$ are often employed, e.g., the shapes of natural oscillations for simple elastic systems (for instance, clamped beams). The modal approach is used for manipulators with elastic links in References 44, 123, 159, 162, 175, 229–235. The approximation error of Equation 4.1.1, depending on the number of terms, is investigated in Reference 234. A comparison of the finite element and modal methods is given for a beam pendulum in Reference 196 and for a two-beam flexible robot arm in Reference 117.

Book[43] considers a mechanical system containing two rigid bodies bound by a multimember linkage. The massless links are connected by joints at which servo drives are located. This system can serve as a model of a space manipulator mounted on a satellite and handling a massive load. The linearized dynamic equations are derived. Vukobratovic and Potkonjak[245] present a numerical algorithm for analysis of oscillations of an elastic manipulator. Some illustrative examples are given.

Numerical algorithms combining the finite element method with the matrix method of dynamical analysis are developed in Reference 221 for spatial mechanisms with elastic links of complex shape. This approach is applied in Reference 222 to manipulation robots with elastic links. The dynamical properties of drives and control system are taken into account. The paper of Sliede et al.[212] is devoted to numerical simulation of small oscillations of an elastic multilink manipulator in a neighborhood of the nominal motion planned for the manipulator with rigid links. The algorithm is based on the finite element method.

In Rakhmanov et al.,[195] differential equations of motion are derived for a four-link manipulator on a moving base. Two links of the manipulator are elastic, whereas its mass is small as compared with masses of the base and a payload. A computer simulation of two-dimensional motions of a two-link elastic manipulator with a massive payload is carried out in the paper of Akulenko et al.[10] The structure of such a manipulator is described in detail in Section 4.2 of this book. The behavior of the system is studied for several typical laws of changing the joint angles. In Zaremba[261] controlled motions of a plane multilink elastic manipulator are analyzed. The links are regarded as inertia-free or modeled by a system of elastically connected mass points. The oscillations of a two-link manipulator are investigated.

Different aspects of mathematical modeling and computer simulation of manipulators with elastic parts are also considered in References 12, 13, 59, 70, 72, 115, 124, 154, 200, 201, 205, 210, 214, 236, 239, 241, 252, and other publications.

The development of effective control algorithms for flexible robots is a challenging problem of modern robotics. This topic is beyond the scope of our book, and we give here only a brief review of some publications. A survey of basic approaches to the problem with the list of relevant publications one can find in Reference 228.

Control laws for a single manipulator link with lumped or distributed elasticity are constructed in References 4, 5, 9, 11, 21, 25, 27, 35, 38, 73, 89, 150–152, 191, 197, 198, 211, 256, etc. In References 5 and 35, open-loop control laws for rotation of an elastic beam are proposed. These laws provide damping a finite number of lower elastic oscillation modes, the residual oscillations being estimated. In Bolotnik and Gukasyan,[38] the algorithm of Akulneko and Bolotnik[5] is applied to a cylindrical coordinate manipulator with an elastic arm. Another approach to the control of such a manipulator is given by Berbyuk and Demidyuk.[26] In Bayo et al.,[21] the method for calculating an open-loop control of a single-link elastic manipulator is developed, which ensures a prescribed motion of an end-effector. The method is based on the inverse dynamics technique and the finite element model of the elastic link. The effectiveness of the proposed approach is corroborated experimentally. This approach is extended to multilink manipulators.[22] In Akulenko and Gukasyan,[9] the control minimizing the integral quadratic functional is proposed for an elastic link.

Linear feedback control laws stabilizing an elastic link at a desired position are designed in References 150–152. The system is controlled by a DC motor. The feedback uses information about the displacements and velocities of the link. A delay in the control circuit is taken into account. The domains of the asymptotic stability in the space of control parameters are found. Lavrovskii and Formalskii[150] deal with bending deformations of a translationally moving beam, longitudinal deformations,[152] and bending deformations of a rotating beam.[151] Feedback controls of rotating elastic beams (manipulator links) are also considered in References 11, 89, and 248. A number of control problems for a single-link electromechanical manipulator with an elastic joint are considered in References 73, 191, 197, 198, and 260. In these papers, a combination of optimal control methods, parametric optimization, and asymptotic techniques is applied.

In Book et al.,[44] different feedback control laws for a two-link elastic manipulator with a point load are proposed and compared with each other. Joint torques are taken as control variables. The paper of Lakota et al.[149] is devoted to the control of a two-link, three-degree-of-freedom manipulator whose mass is much less than the mass of a payload. The manipulator is equipped with DC drives. The feedback control laws moving the load along a given straight line are presented. The stability of the motion is studied. A manipulator of similar kinematic structure is considered by Pfeiffer et al.[189] Unlike Lakota et al.,[149] the mass of the load is assumed to be comparable with that of the manipulator, and both the links and joints are elastic. Joint torques are taken as control functions. The control is proposed which provides effective damping of elastic oscillations in a vicinity of a prescribed reference trajectory. The control law combines the feedforward and feedback modes. Theoretical results are compared with experimental data.

In Reference 95, for a flexible gantry robot, the force feedback control is described which ensures contact with a manipulation object. The influence of the feedback gains on the stability of the motion is investigated. The paper of Becker et al.[23] deals with precise control of a robot for machining metal workpieces of a complex shape. The control law is designed that provides the motion of a tool along a prescribed trajectory. The computation of control functions is based on the mathematical model, taking into account the elasticity of the manipulator structure, dry friction, and backlashes in gear trains.

Various control laws involving active damping of elastic oscillations in manipulation robots are developed and analyzed by Eliseyev et al.[85] The book of Berbyuk[25] contains some control algorithms for an elastic robot arm. These algorithms are based on a combination of inverse dynamics and parameter optimization techniques. The approach to control algorithms for flexible manipulators based on optimal control methods and asymptotic methods of nonlinear mechanics is described by Akulenko.[3]

Different aspects of control of manipulation robots with elastic parts are also discussed in References 14, 45, 52, 72, 76, 91, 128–130, 174, 200, 201, 225, 229, 230, 234, 235, 257, and other publications.

In subsequent Sections of Chapter 4, we deal with mathematical simulation of the dynamics of a manipulator with flexible links. Note that flexibility of links is essential, in particular, for large space manipulators.

Chapter 5 is devoted to experimental investigation of elastic properties of industrial robots. It is shown that the elastic compliance of these robots is due mainly to the flexibility of joints. In this connection, in Chapter 6 we study the dynamics of industrial manipulation robots with rigid links and elastic joints.[101]

In the majority of publications, including those mentioned above, a computer simulation of the dynamics of flexible robots is widely used. Differential equations of motion are integrated straightforwardly by well-known numerical methods. The integration step must be short, since the stiffness of links and/or joints of robots, as a rule, is great, and therefore the elastic oscillations have brief periods compared with the characteristic duration of the motion. Thus, the simulation requires much computational time and is accompanied by the accumulation of round-off errors.

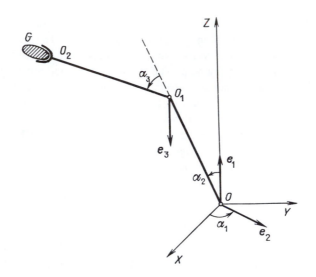

FIGURE 4.2.1.

The distinctive feature of our approach is the combination of asymptotic and numerical techniques. In accordance with the general method of Chapter 3, we represent the whole motion as a superposition of a slow component and fast elastic oscillations. Equations of the fast oscillations are solved in the explicit form, whereas equations of the slow motion are integrated numerically with a comparatively long step. This approach reduces the computation time considerably.

4.2. MANIPULATOR WITH ELASTIC LINKS

4.2.1. INITIAL ASSUMPTIONS
In this chapter we consider a manipulation robot with elastic links under the following assumptions:

1. A manipulator OO_1O_2 (Figure 4.2.1) consists of two elastic links (beams) of equal length l.
2. At the fixed point O, two revolute joints are combined. The axis $\mathbf{e}_1$ of the first joint is fixed, while the axis $\mathbf{e}_2$ of the second joint can rotate about the axis of the first joint and is orthogonal to $\mathbf{e}_1$. At the point O_1, the revolute elbow joint is placed; its axis is parallel to $\mathbf{e}_2$ if the link OO_1 is not deformed. At the point O_2, a spherical joint is placed, i.e., three revolute joints are combined here.
3. A payload G is a rigid body of the mass m.
4. The manipulator links are homogeneous rectilinear beams of annular cross section which can undergo bending and torsion deformations.

5.   Elastic displacements of the manipulator and the payload during the motion are small compared with the length l of the links.
6.   The manipulator mass m_0 is small compared with the load mass m.
7.   Only control torques in the joints and the reaction force at the base O act upon the manipulator; other forces are not taken into account.

Let us discuss these assumptions.

Assumptions 1 and 2 are idealizations of the real kinematics of an anthropomorphic manipulator. We ignore the dimensions of the hand and the distance between the joint axes e_1 and e_2. Usually, these dimensions are small in comparison with the length l of the upper arm and forearm and therefore do not influence significantly elastic oscillations which are mostly caused by upper arm and forearm rotations.

Assumptions 4 and 5 are usually satisfied for real manipulators.

Assumption 6 is true for space manipulators which are comparatively light and carry a large payload in zero gravity conditions. This assumption implies specific properties of the elastic oscillations spectrum. Denote by c the characteristic stiffness of the manipulator structure. Then the natural frequencies of elastic oscillations are separated into two groups: the frequencies of the order $(c/m)^{1/2}$ and those of the order $(c/m_0)^{1/2}$. The frequencies of the first group (low) correspond to quasistatic deformations of the manipulator. Their number is equal to the number of degrees of freedom of the load G. In the general case, this number is equal to six. However, if linear dimensions of the load G are small compared with the length l, the load can be regarded as a point mass, and then the number of low frequencies is three.

The frequencies of the second group (high) correspond to elastic waves in the beams OO_1 and O_1O_2; their spectrum is infinite. The oscillations of the second group correspond to small displacements of the load in comparison with displacements of points of the manipulator. These high-frequency oscillations decay faster than oscillations of the first group. Therefore, we can neglect the oscillations of the second group and take into account only the oscillations of the first group. For these oscillations, elastic deformations of the links can be considered as quasistatic. In other words, assumption 6 implies that at any time instant the elastic manipulator OO_1O_2 has an equilibrium configuration.

Assumptions 4 and 5 allow us to use the linear theory of elastic beam deformations. In this theory, extension deformations are small compared with bending and torsion deformations. Therefore, with high accuracy the beams can be regarded as inextensible. Note that assumption 5 can also be stated in the following equivalent form:

5′. The periods of natural elastic oscillations of the manipulator are small in comparison with the characteristic time of a manipulation operation.

Indeed, denote by T_0 the characteristic time of the operation. Then the characteristic acceleration of the load is lT_0^{-2} and the characteristic force applied to the load by the manipulator is $\Phi \sim mlT_0^{-2}$. The order of magnitude of the elastic displacement u of load is given by

$$u \sim \Phi c^{-1} = mlT_0^{-2}c^{-1} \tag{4.2.1}$$

Denoting by $T_1 \sim (m/c)^{1/2}$ the period of the lowest elastic oscillations mode, we obtain from Equation 4.2.1:

$$u/l \sim (T_1/T_0)^2 \ll 1 \tag{4.2.2}$$

Equation 4.2.2 proves the equivalence of assumptions 5 and 5′.

Assumption 5′ means that the angular velocities of link rotation are small compared with the lowest elastic oscillation frequency. Therefore, we can ignore the influence of centrifugal forces on the manipulator stiffness and determine the stiffness matrix according to elastic beams statics.

Note that Equation 4.2.2 allows us to introduce a small parameter, $\epsilon - T_1/T_0 \ll 1$, and to use an asymptotic approach.

Assumption 7 does not restrict our approach and is made for simplicity. The influence of different external forces can be readily taken into account.

Decay of elastic oscillations caused by energy dissipation in the robot links and joints is not considered here. Since the decay evidently leads to reducing the elastic oscillation amplitudes, the results obtained below estimate the elastic displacements from above.

As in Chapter 3, a manipulator whose links OO_1 and O_1O_2 are rigid is called a rigid model.

Denote by α_1 the angle of rotation of the manipulator about the joint axis e_1, by α_2 the angle of rotation of the link OO_1 about the joint axis e_2 (this angle is measured from the axis e_1), and by α_3 the rotation angle at the elbow joint (see Figure 4.2.1). Let us introduce the inertial Cartesian reference frame whose Z axis is pointed along the fixed joint axis e_1. The moving joint axis e_2 lies in the OXY plane and forms the angle α_1 with respect to the axis X. The angle between the tangent to the link OO_1 at the point O and the axis Z is α_2, while the angle between the tangents to the links OO_1 and O_1O_2 at the point O_1 is α_3. We denote by α_4, α_5, and α_6 the angles of rotations of the load G about three axes of the revolute joints placed at the endpoint O_2 of the second link.

4.2.2. POTENTIAL ENERGY

Let us determine the elastic potential energy of the manipulator. First, we give the formula for the potential energy of a rectilinear homogeneous elastic beam of the length l. Let the beam be rigidly fixed (as a cantilever) at its one (left) end, while for the other end, the vectors of elastic displacement, u, and twist, θ, with respect to the undeformed state, be specified. Denote by $w(s)$ the vector of elastic displacement of the beam neutral line and by

$\chi(s)$ the twist angle of the beam cross section about its axis. Here, the argument s is the length of the beam measured from its left end, $0 \leq s \leq l$. The potential elastic energy of the deformed beam is given by (see, for instance, Reference 158):

$$\Pi = \frac{EI}{2} \int_0^l [\mathbf{w}''(s)]^2 ds + \frac{C}{2} \int_0^l [\chi'(s)]^2 ds \qquad (4.2.3)$$

Here, E is Young's modulus of the beam material, I is the inertia moment of the beam cross section with respect to its diameter, and C is the beam torsional stiffness. Primes mean the derivatives with respect to s. We have

$$I = \frac{\pi(\rho_2^4 - \rho_1^4)}{4}, \; C = \frac{E\pi(\rho_2^4 - \rho_1^4)}{4(1 + \nu)}$$

for the beam of the annular cross section whose inner and outer radii are ρ_1 and ρ_2, respectively; ν is the Poisson coefficient.

The functions $\mathbf{w}(s)$, $\chi(s)$ in the quasistatic (equilibrium) state satisfy the equation and boundary conditions,

$$\left| \mathbf{w}^{\mathrm{Iv}}(s) = 0, \chi''(s) = 0, 0 \leq s \leq l, \right|$$
$$\left| \mathbf{w}(0) = \mathbf{w}'0) = 0, \chi(0) = 0, \right|$$
$$\left| \mathbf{w}(l) = u, \mathbf{w}'(l) = \mathbf{\theta} \times \mathbf{k}, \chi(l) = \mathbf{\theta} \cdot \mathbf{k} \right| \qquad (4.2.4)$$

where $\mathbf{k}$ is a unit vector directed along the axis of the beam when it is undeformed. The boundary conditions in Equations 4.2.4 for the left end correspond to the rigid fixation, while the conditions for the right end reflect the fact that the vectors of elastic displacement and rotation are $\mathbf{u}$ and $\mathbf{\theta}$, respectively. Indeed, the unit vector of the tangent to the beam at its right end is

$$\mathbf{\tau} = \mathbf{r}'(l) = d(\mathbf{k}s + \mathbf{w})/ds = \mathbf{k} + \mathbf{w}'(l)$$

On the other hand, the unit vector of the beam axis can be expressed as

$$\mathbf{\tau} = \mathbf{k} + \mathbf{\theta} \times \mathbf{k}$$

The obtained relationships entail the boundary condition $\mathbf{w}'(l) = \mathbf{\theta} \times \mathbf{k}$ in Equations 4.2.4. The last condition in Equations 4.2.4 means that the twist angle at the right end of the beam is fixed.

The solution of the boundary value problem (Equations 4.2.4) is given by

$$\mathbf{w}(s) = (3\mathbf{u} - l\mathbf{\theta} \times \mathbf{k})(s/l)^2 + (l\mathbf{\theta} \times \mathbf{k} - 2\mathbf{u})(s/l)^3$$
$$\chi(s) = \mathbf{\theta} \cdot \mathbf{k}(s/l) \qquad (4.2.5)$$

Substituting $\mathbf{w}\,(s)$ and $x\,(s)$ from Equations 4.2.5 into Equation 4.2.3, we obtain the potential energy of the beam:

$$\Pi = \frac{2EI}{l^3}\,\{3\mathbf{u}^2 - 3\mathbf{u}\cdot(\boldsymbol{\theta} \times \mathbf{k})l + l^2(\boldsymbol{\theta} \times \mathbf{k})^2\} + \frac{C}{2l}\,(\boldsymbol{\theta}\cdot\mathbf{k})^2 \quad (4.2.6)$$

Let us now proceed to the elastic two-link manipulator, OO_1O_2, both of whose links are beams of annular or circular cross section. Denote by E_jI_j and C_j the bending and torsional stiffnesses of the beams, respectively; here, the indices $j = 1, 2$ refer to the links OO_1 and O_1O_2, respectively.

Denote by $\mathbf{k}_j$ the unit vector directed along the axis of the jth undeformed link and by $\mathbf{u}_j,\ \boldsymbol{\theta}_j$ the vectors of displacement and twist of the end of the jth link with respect to the undeformed configuration of the manipulator, $j = 1, 2$. The displacement of the end of the second link (the point O_2), due to elastic deformation of the first link, is $\mathbf{u}_2^0 = \mathbf{u}_1 + \boldsymbol{\theta}_1 \times \mathbf{k}_2 l$. The vectors of displacement and twist of the end of the second link, due to its own deformation, are expressed as

$$\mathbf{u}_2 - \mathbf{u}_2^0 = \mathbf{u}_2 - \mathbf{u}_1 - \boldsymbol{\theta}_1 \times \mathbf{k}_2 l,\ \boldsymbol{\theta}_2 - \boldsymbol{\theta}_1 \quad (4.2.7)$$

respectively.

Let us determine the potential energy of the two-link manipulator using Equation 4.2.6 and substituting for $\mathbf{u},\ \boldsymbol{\theta}$ the vectors $\mathbf{u}_1,\ \boldsymbol{\theta}_1$ for the first link, and the vectors from Equation 4.2.7 for the second link. We obtain

$$\Pi = \Pi_1 + \Pi_2 = \frac{2E_1I_1}{l^3}\,[3\mathbf{u}_1^2 - 3\mathbf{u}_1\cdot(\boldsymbol{\theta} \times \mathbf{k}_1)l + (\boldsymbol{\theta}_1 \times \mathbf{k}_1 l)^2]$$

$$+ \frac{C_1}{2l}\,(\boldsymbol{\theta}_1\cdot\mathbf{k}_1)^2 + \frac{2E_2I_2}{l^3}\,\{3(\mathbf{u}_2 - \mathbf{u}_1 - \boldsymbol{\theta}_1 \times \mathbf{k}_2 l)^2$$

$$- 3(\mathbf{u}_2 - \mathbf{u}_1 - \boldsymbol{\theta}_1 \times \mathbf{k}_2 l)\cdot[(\boldsymbol{\theta}_2 - \boldsymbol{\theta}_1) \times \mathbf{k}_2 l]$$

$$+ [(\boldsymbol{\theta}_2 - \boldsymbol{\theta}_1) \times \mathbf{k}_2 l]^2\} + \frac{C_2}{2l}\,[(\boldsymbol{\theta}_2 - \boldsymbol{\theta}_1)\cdot\mathbf{k}_2]^2 \quad (4.2.8)$$

Let us write down the inextensibility conditions of the links as

$$\mathbf{u}_1\cdot\mathbf{k}_1 = 0,\ (\mathbf{u}_2 - \mathbf{u}_1)\cdot\mathbf{k}_2 = 0 \quad (4.2.9)$$

We connect the Cartesian reference frame $O_2X'Y'Z'$ with the link O_1O_2 of the rigid model. The axis Z' is directed along the undeformed link O_1O_2, the axis X' coincides with the axis of the joint O_1, and the axis Y' complements the system to the right-hand orthogonal one.

Let us project the vectors $\mathbf{u}_j$, $\mathbf{\theta}_j$, $j = 1,2$, onto the axes of the reference frame $O_2X'Y'Z'$ and introduce the denotations

$$\mathbf{u}_j = \|x_j, y_j, z_j\|, \ \mathbf{\theta}_j = \|\varphi_j, \psi_j, \chi_j\| \tag{4.2.10}$$

for the respective components of the vectors $\mathbf{u}_j$ and $\mathbf{\theta}_j$.

The components of the vectors $\mathbf{k}_1$ and $\mathbf{k}_2$ in this reference frame are given by

$$\mathbf{k}_1 = \|0, \sin \alpha_3, \cos \alpha_3\|, \ \mathbf{k}_2 = \|0, \ 0, \ 1\| \tag{4.2.11}$$

Taking account of Equations 4.2.10 and 4.2.11, we can rewrite Equations 4.2.9 as follows:

$$y_1\sin \alpha_3 + z_1\cos \alpha_3 = 0, \ z_2 - z_1 = 0 \tag{4.2.12}$$

From Equations 4.2.12 we express y_1, z_1 through z_2, α_3 and substitute $y_1 = -z_2\cot\alpha_3$, $z_1 = z_2$ into the expression for elastic potential energy (Equation 4.2.8). Then the potential energy becomes a quadratic form of ten independent variables: x_1 and nine components of the vectors $\mathbf{u}_2$, $\mathbf{\theta}_1$, and $\mathbf{\theta}_2$. Coefficients of the quadratic form depend on the angle α_3 between the links OO_1 and O_1O_2. This angle determines the configuration of the elastic manipulator.

It is convenient to introduce the following vectors and scalars:

$$v_i' = \|x_i, \psi_i, \chi_i\|^T, \ i = 1,2$$

$$v_1'' = \varphi_1 \tag{4.2.13}$$

$$v_2'' = \|y_2, z_2, \varphi_2\|^T$$

Now we rewrite potential energy (Equation 4.2.8) using variables given by Equation 4.2.13:

$$\Pi = \frac{1}{2} \sum_{i,j=1}^{2} [(C_{ij}'v_i', v_j') + C_{ij}''v_i'', v_j'')] \tag{4.2.14}$$

Here, C_{ij}' and C_{22}'' are symmetric 3×3 matrices, C_{11}'' is a scalar, and C_{12}'' is a column matrix of the dimension 1×3, $C_{21}'' = (C_{12}'')^T$. These matrices are given by

$$C'_{11} = \begin{Vmatrix} 12(a_1 + a_2)l^{-3} & 6(a_2 - a_1c)l^{-2} & 6a_1sl^{-2} \\ 6(a_2 - a_1c)l^{-2} & (4a_1c^2 + C_1s^2 + (4a_2)l^{-1} & -(4a_2 - C_1)scl^{-1} \\ 6a_1sl^{-2} & -(4a_2 - C_1)scl^{-1} & (4a_1s^2 + C_1c^2 + 4a_2)l^{-1} \end{Vmatrix}$$

$$C'_{12} = \begin{Vmatrix} -12a_2l^{-3} & 6a_2l^{-1} & 0 \\ 6a_2l^{-1} & 2a_2l^{-1} & 0 \\ 0 & 0 & -C_2l^{-1} \end{Vmatrix}$$

$$C'_{21} = C'_{12}$$

$$C'_{22} = \begin{Vmatrix} 12a_2l^{-3} & -6a_2l^{-1} & 0 \\ -6a_2l^{-1} & 4a_2l^{-1} & 0 \\ 0 & 0 & -C_2l^{-1} \end{Vmatrix}$$

$$C''_{11} = 4(a_1 + a_2)l^{-1}, \quad C''_{12} = \|6a_2l^{-2}, 6(-a_1 + a_2c)sl^{-2}, 2a_2l^{-1}\|^T$$

$$C''_{21} = (C''_{12})^T$$

$$C''_{22} = \begin{Vmatrix} 12a_2l^{-3} & 12a_2l^{-3}\tan^{-1}\alpha_3 & 6a_2l^{-2} \\ 12a_2l^{-3}\tan^{-1}\alpha_3 & 12(a_1 + a_2c^2)l^{-3}s^{-2} & 6a_2l^{-2}\tan^{-1}\alpha_3 \\ 6a_2l^{-2} & 6a_2l^{-2}\tan^{-1}\alpha_3 & 4a_2l^{-1} \end{Vmatrix}$$

$$a_i = E_iI_i, \ i = 1,2; \ s = \sin\alpha_3, \ c = \cos\alpha_3 \tag{4.2.15}$$

By C_j in Equations 4.2.15 we have denoted the torsional stiffness of the jth link of the manipulator, $j = 1,2$.

Note that in Equation 4.2.14 for the elastic potential energy, we can distinguish two independent groups of terms. The first group includes the vector v'_i and matrices C'_{ij}, while the other comprises the vectors v''_i and matrices C''_{ij}, $i, j = 1,2$. Components of the vectors v'_1 and v'_2 determine the deformation of the manipulator OO_1O_2 in the direction normal to the plane of the undeformed manipulator, while components of the vectors v''_1, v''_2 describe the deformation in this plane. The quadratic form (Equation 4.2.14) depends on ten variables (Equations 4.2.13). However, for an arbitrary equilibrium state of the elastic manipulator OO_1O_2, only six variables can be specified independently, viz., the vectors v'_2, v''_2 describing the deformation at the end O_2 of the manipulator. The variables v'_1 and v''_1 in Equations 4.2.13 must take the equilibrium values, corresponding to the minimum of potential energy (Equation 4.2.14) over these variables. Thus, v'_1 and v''_1 are determined by the conditions

$$\partial\Pi/\partial v'_1 = \partial\Pi/\partial v''_1 = 0 \tag{4.2.16}$$

Substituting the expression for the elastic potential energy (Equation 4.1.14) into Equation 4.2.16, we obtain

$$C'_{11}v'_1 + C'_{21}v'_2 = 0, \quad C''_{11}v''_1 + C''_{21}v''_2 = 0 \tag{4.2.17}$$

Let us express the vectors v_1' and v_1'' from Equations 4.2.17:

$$v_1' = -(C_{11}')^{-1}C_{21}'v_2', \quad v_1'' = -(C_{11}'')^{-1}C_{21}''v_2''$$

and substitute these relationships into Equation 4.2.14. Then, having reduced like terms, we obtain

$$\Pi = \frac{1}{2}[(D'v_2', v_2') + (D''v_2''v_2'')]$$

$$D' = C_{22}' - (C_{21}')^{\mathrm{T}}(C_{11}')^{-1}C_{21}'$$

$$D'' = C_{22}'' - (C_{21}'')^{\mathrm{T}}(C_{11}'')^{-1}C_{21}'' \tag{4.2.18}$$

For further analysis, it is convenient to combine the vectors v_2' and v_2'' in one vector $v^{(2)}$ and write down potential energy (Equation 4.2.18) as

$$\Pi = \frac{1}{2}(C_2 v^{(2)}, v^{(2)})$$

$$v^{(2)} = \|x_2, \psi_2, \chi_2, y_2, z_2, \varphi_2\|$$

$$C_2 = \begin{Vmatrix} D' & 0 \\ 0 & D'' \end{Vmatrix} \tag{4.2.19}$$

where D' and D'' are defined by Equations 4.2.18 and 0 is the zero 3×3 matrix.

Thus, we have represented the elastic potential energy of the manipulator OO_1O_2 as the quadratic form (Equation 4.2.19), depending on elastic displacement $\mathbf{u}_2 = \|x_2, y_2, z_2\|$ and twist $\boldsymbol{\theta}_2 = \|\varphi_2, \psi_2, \chi_2\|$ at the end of the second link (see Equations 4.2.10). Coefficients of the quadratic form in Equation 4.2.19 depend on the angle α_3 between the links OO_1 and O_1O_2, as well as on the elastic parameters of the links.

Denote by $\boldsymbol{\Phi}$ the resultant vector of forces applied to the manipulator by the load and by $\boldsymbol{\mu}$ the torque of these forces with respect to the point O_2. These forces are balanced by elastic forces acting on the load. Their resultant vector and torque with respect to the point O_2 are $-\partial\Pi/\partial\mathbf{u}_2$ and $-\partial\Pi/\partial\boldsymbol{\theta}$, respectively. Therefore,

$$\boldsymbol{\Phi} - \frac{\partial\Pi}{\partial\mathbf{u}_2} = 0, \quad \boldsymbol{\mu} - \frac{\partial\Pi}{\partial\boldsymbol{\theta}_2} = 0 \tag{4.2.20}$$

Let us consider now the special case which is repeatedly met in what follows. Suppose the load G is a point mass, its dimensions and moments of inertia being neglected. Then the torque $\boldsymbol{\mu}$ is equal to zero. In this case the twist vector $\boldsymbol{\theta}_2 = \|\varphi_2, \psi_2, \chi_2\|$ (see Equations 4.2.10), is determined from equations 4.2.20 with $\boldsymbol{v} = 0$, i.e.,

$$\frac{\partial \Pi}{\partial \varphi_2} = \frac{\partial \Pi}{\partial \psi_2} = \frac{\partial \Pi}{\partial \chi_2} = 0 \tag{4.2.21}$$

Substituting the expression for the elastic potential energy (Equation 4.2.19) into Equations 4.2.21 we arrive at the linear equations,

$$\begin{cases} d'_{21}x_2 + d'_{22}\psi_2 + d'_{23}\chi_2 = 0 \\ d'_{31}x_2 + d'_{32}\psi_2 + d'_{33}\chi_2 = 0 \end{cases} \tag{4.2.22}$$

$$d''_{31}y_2 + d''_{32}z_2 + d''_{23}\varphi_2 = 0 \tag{4.2.23}$$

where d'_{ij} and d''_{ij}, $i,j = 1,2,3$ are elements of the matrices D' and D'', respectively. Using Equation 4.2.23 we express φ_2 through y_2 and z_2, and then find ψ_2 and χ_2 from linear Equations 4.2.22. It turns out that $\chi_2 = 0$, i.e., the torsion of the second link O_1O_2 is absent. Inserting obtained expressions for φ_2, ψ_2, and χ_2 into Equation 4.2.19 and taking account of Equations 4.2.15 and 4.2.18, we have

$$\Pi = \frac{1}{2}(C_0 \mathbf{u}_2, \mathbf{u}_2)$$

$$\mathbf{u}_2 = (x_2, y_2, z_2), \quad C_0 = k\|c_{ij}\|, \quad i,j = 1, 2, 3$$

$$c_{11} = (4\xi + 3)(3\zeta^{-1}\sin^2 \alpha_3 + \xi + 3\cos \alpha_3 + 3\cos^2 \alpha_3 + 1)^{-1}$$

$$c_{12} = c_{21} = c_{13} = c_{31} = 0, \quad c_{22} = 4$$

$$c_{23} = c_{32} = -2(3 + 2\cos \alpha_3)/\sin \alpha_3$$

$$c_{33} = 4(3 + \xi + 3\cos \alpha_3 + \cos^2 \alpha_3)/\sin^2 \alpha_3$$

$$k = 3E_1 I_1 l^{-3}(4\xi + 3)^{-1}, \quad \xi = E_1 I_1/(E_2 I_2), \quad \zeta = C_1/(E_2 I_2) \tag{4.2.24}$$

Thus, in the general case, the matrix of stiffness is expressed through C'_{ij} and C''_{ij}, $i, j = 1, 2$ (see Equations 4.2.19, 4.2.18, and 4.2.15). In a particular case, when the torque at the manipulator gripper is equal to zero, elements of the stiffness matrix are given by explicit formulas in Equations 4.2.24. Note that the matrices C'_{ij}, C''_{ij}, and C_0 are presented in the reference frame $O_2X'Y'Z'$ in which they have the simplest form (elements of the matrices depend only on the angle α_3).

To determine the kinetic energy, it is convenient to introduce the Cartesian reference frame $O_cX''Y''X''$ with the origin at the mass center O_c of the load

G and axes directed along the principal inertia axes of the load (for the undeformed manipulator). Thus, the orientation of the frame $O_cX''Y''Z''$ depends only on the joint angles $\alpha_1, \ldots, \alpha_6$.

Similar to $\mathbf{u}_i$, $\boldsymbol{\theta}_i$, we denote by $\mathbf{u}$ the vector of elastic displacement of the mass center O_c of the load G, and by $\boldsymbol{\theta}$ the vector of elastic twist of the body G. Denote the components of these vectors in the reference frame $O_cX''Y''Z''$ as follows,

$$\mathbf{u} = \|x, y, z\|, \; \boldsymbol{\theta} = \|\varphi, \psi, \chi\| \tag{4.2.25}$$

and introduce the six-dimensional vector:

$$v = \|x, \psi, \chi, y, z, \varphi\| \tag{4.2.26}$$

Let us determine the matrix W connecting the vector $v^{(2)}$ from Equation 4.2.19 and the vector v from Equation 4.2.26:

$$v^{(2)} = Wv \tag{4.2.27}$$

The vectors $\mathbf{u}$, $\boldsymbol{\theta}$, $\mathbf{u}_2$, $\boldsymbol{\theta}_2$, and $\overrightarrow{O_2O_c}$ are connected by the kinematic relations

$$\mathbf{u}_2 = \mathbf{u} - \boldsymbol{\theta} \times \overrightarrow{O_2O_c}, \; \boldsymbol{\theta}_2 = \boldsymbol{\theta} \tag{4.2.28}$$

Let l_x, l_y, and l_z be components of the vector $\overrightarrow{O_2O_c}$ in the reference frame $O_2X'Y'Z'$:

$$\overrightarrow{O_2O_c} = \|l_x, l_y, l_z\|$$

Denote by $V = \|v_{ij}\|$ the matrix of direction cosines for the axes of the reference frame $O_cX''Y''Z''$ with respect to the frame $O_2X'Y'Z'$. This matrix is given by the relationship similar to Equation 2.2.2:

$$V = \|v_{ij}\| = \begin{Vmatrix} \cos(X'', X') & \cos(Y'', X') & \cos(Z'', X') \\ \cos(X'', Y') & \cos(Y'', Y') & \cos(Z'', Y') \\ \cos(X'', Z') & \cos(Y'', Z') & \cos(Z'', Z') \end{Vmatrix} \tag{4.2.29}$$

From Equations 4.2.28 and 4.2.29, the equalities stem connecting the components of the vectors $\mathbf{u}_2$, $\boldsymbol{\theta}_2$, $\mathbf{u}$, and $\boldsymbol{\theta}$ (Equations 4.2.10 and 4.2.25) in the corresponding reference frames

$$\{\mathbf{u}_2\}' = \{\mathbf{u}\}' - \{\mathbf{\theta}\}' \times \{\overrightarrow{O_2O_c}\}' = V\{\mathbf{u}\}'' - (V\{\mathbf{\theta}\}'') \times \{\overrightarrow{O_2O_c}\}'$$

$$\{\mathbf{\theta}\}' = V\{\mathbf{\theta}\}'' \qquad\qquad (4.2.30)$$

The superscripts $'$ and $''$ indicate that the corresponding vectors are presented in the reference frames $O_2X'Y'Z'$ and $O_cX''Y''Z''$, respectively. Equations 4.2.30 represent the explicit form of Equation 4.2.27.

The matrix W from Equation 4.2.27 is expressed by

$$W = W_1 W_2$$

$$W_1 = \begin{Vmatrix} 1 & -l_z & l_y & 0 & 0 & 0 \\ 0 & 1 & 0 & 0 & 0 & 0 \\ 0 & 0 & 1 & 0 & 0 & 0 \\ 0 & 0 & -l_x & 1 & 0 & l_z \\ 0 & l_x & 0 & 0 & 1 & -l_y \\ 0 & 0 & 0 & 0 & 0 & 1 \end{Vmatrix}$$

$$W_2 = \begin{Vmatrix} v_{11} & 0 & 0 & v_{12} & v_{13} & 0 \\ 0 & v_{22} & v_{23} & 0 & 0 & v_{21} \\ 0 & v_{32} & v_{33} & 0 & 0 & v_{31} \\ v_{21} & 0 & 0 & v_{22} & v_{23} & 0 \\ v_{31} & 0 & 0 & v_{32} & v_{33} & 0 \\ 0 & v_{12} & v_{13} & 0 & 0 & v_{11} \end{Vmatrix} \qquad (4.2.31)$$

Substituting Equation 4.2.27 into Equation 4.2.19, we obtain the potential energy as a function of the vector v:

$$\Pi = \frac{1}{2}(Cv, v), \quad C = W^T C_2 W$$

$$v = \|x, \psi, \chi, y, z, \varphi\| \qquad\qquad (4.2.32)$$

4.3. FREE OSCILLATIONS OF ELASTIC MANIPULATOR

4.3.1. EIGENVALUE PROBLEM

In Section 4.3 we analyze free elastic oscillation of the two-link manipulator described in the previous section. Natural frequencies and shapes of the elastic oscillations are calculated. These results are subsequently used when investigating controlled motions of the manipulator. Besides, they are of interest by themselves.

Throughout this section, we assume all joint angles $\alpha_1, \ldots, \alpha_6$ to be fixed. Due to the assumption 6 of Section 4.2, the kinetic energy of the system is reduced to the kinetic energy of the load. In the case of small elastic displacements and twists, the kinetic energy is given by

$$T = \frac{1}{2}(A\dot{v}, \dot{v}), \quad A = \text{diag}\{m, J_2, J_3, m, m, J_1\} \tag{4.3.1}$$

Here, v is given by Equation 4.2.26, m is the load mass; J_1, J_2, and J_3 are the principal moments of inertia of the load at its mass center. Let us write down Lagrangian Equations 2.3.11 for the manipulator with a load using Equations 4.2.32 for the potential energy, and Equation 4.3.1 for the kinetic energy. We obtain the set of six linear differential equations:

$$A\ddot{v} + Cv = 0 \tag{4.3.2}$$

The general solution of Equation 4.3.2 can be presented as a sum of partial solutions of the form

$$v(t) = b\sin(\omega t + \delta) \tag{4.3.3}$$

Equation 4.3.3 describes free elastic oscillations of the manipulator, ω, b, and δ being the frequency, amplitude vector, and initial phase of the oscillations, respectively. Substituting Equation 4.3.3 into Equation 4.3.2 we obtain the set of linear equations for components of the vector b:

$$(C - \omega^2 A)b = 0 \tag{4.3.4}$$

Having equated the determinant of this system to zero, we get the characteristic equation

$$\det(C - \omega^2 A) = 0 \tag{4.3.5}$$

Since the matrices A and C are positive definite, the roots ω_i^2, $i = 1, \ldots, 6$, of Equation 4.3.5 are real and positive. The quantities ω_i are natural frequencies of elastic oscillations of the system. Having determined ω_i, $i = 1, \ldots, 6$, we substitute them into Equation 4.3.4 and find amplitude vectors b^i, $i = 1, \ldots, 6$. Note that there may be multiple roots, and we will count each root ω_i according to its multiplicity. However, there are always six linearly independent amplitude vectors satisfying linear algebraic Equation 4.3.4. The general solution of Equation 4.3.2 can be expressed as

$$v(t) = \sum_{i=1}^{6} a_i b^i \sin(\omega_i t + \delta_i) \tag{4.3.6}$$

Here, a_i and δ_i are arbitrary constants determined by initial conditions. In the next section we calculate eigenfrequencies and shapes for some particular cases.

4.3.2. NATURAL MODES OF OSCILLATIONS

Suppose the center of mass of the load for the undeformed manipulator lies in the plane of the manipulator OO_1O_2, and one of the principal inertia axes, O_xX'', is parallel to the axis O_2X' of the reference frame $O_2X'Y'Z'$, i.e., to the elbow joint axis $\mathbf{e}_3$ (see Figure 4.2.1). The two other inertia axes can be arbitrarily orientated in the plane $O_2Y'Z'$. In this case, we obtain that

$$l_x = 0, \; v_{11} = 1, \; v_{12} = v_{21} = v_{13} = v_{31} = 0$$

in Equations 4.2.29 and 4.2.31. Therefore, the matrices W_1 and W_2 become block-diagonal:

$$W_1 = \begin{Vmatrix} W_1' & 0 \\ 0 & W_1'' \end{Vmatrix}, \; W_2 = \begin{Vmatrix} W_2' & 0 \\ 0 & W_2'' \end{Vmatrix}$$

$$W_1' = \begin{Vmatrix} 1 & -l_z & l_y \\ 0 & 1 & 0 \\ 0 & 0 & 1 \end{Vmatrix}, \; W_1'' = \begin{Vmatrix} 1 & 0 & l_z \\ 0 & 1 & -l_y \\ 0 & 0 & 1 \end{Vmatrix}$$

$$W_2' = \begin{Vmatrix} 1 & 0 & 0 \\ 0 & v_{22} & v_{23} \\ 0 & v_{32} & v_{33} \end{Vmatrix}, \; W_2'' = \begin{Vmatrix} v_{22} & v_{23} & 0 \\ v_{32} & v_{33} & 0 \\ 0 & 0 & 1 \end{Vmatrix} \tag{4.3.7}$$

Substituting W_1 and W_2 Equations 4.3.7 into Equations 4.2.31 and 4.2.32, and taking into account Equation 4.2.19 for C_2, we obtain

$$C = \begin{Vmatrix} C' & 0 \\ 0 & C'' \end{Vmatrix}, \; C' = (W')^{\mathsf{T}}D'W'$$

$$C'' = (W'')^{\mathsf{T}}D''W'', \; W' = W_1'W_2', \; W'' = W_1''W_2''$$

The matrices D' and D'' are defined by Equations 4.2.18. Let us substitute the obtained matrix C into characteristic Equation 4.3.5. Since the determinant of a block-diagonal matrix is equal to the product of determinants of the diagonal blocks, we can rewrite Equation 4.3.5 as follows:

$$\det(C - \omega^2 A) = \det(C' - \omega^2 A') \det(C'' - \omega^2 A'') = 0$$

$$A' = \mathrm{diag}\{m, J_2, J_3\}, \; A'' = \mathrm{diag}\{m, m, J_1\}$$

Thus, the spectrum of eigenfrequences separates into two independent groups. The first group includes three modes in which points of the manipulator oscillate normally to the plane of the manipulator OO_1O_2. The second group also includes three frequencies, and the respective oscillations occur in the plane of the manipulator.

Suppose now that the linear dimensions of the load are small compared with the length l of the manipulator links. Then we can regard the load as a point mass placed at the gripper O_2. In this case, the mechanical system has three degrees of freedom, its stiffness matrix C_0 is determined by Equations 4.2.24 and the matrix of the kinetic energy is given by

$$A_0 = \text{diag}\{m, m, m\} \tag{4.3.8}$$

Let us put $A = A_0$ and $C = C_0$ (from Equations 4.3.8 and 4.2.24) in characteristic Equation 4.3.5 and determine the eigenfrequencies:

$$\omega_1^2 = \frac{3E_1I_1}{ml^3} (3\xi\zeta^{-1} \sin^2 \alpha_3 + \xi + 3\cos^2 \alpha_3 + 3\cos \alpha_3 + 1)^{-1}$$

$$\omega_{2,3}^2 = \frac{6E_1I_1}{ml^3} \{\xi + 4 + 3\cos \alpha_3 \pm [P + (3 + 2\cos^2 \alpha_3)\sin^2 \alpha_3]^{1/2}\}^{-1}$$

$$P = (3 + \xi + 3\cos \alpha_3 + \cos 2\alpha_3)^2 \tag{4.3.9}$$

Substituting $\omega^2 = \omega_i^2$ from Equation 4.3.9 into Equation 4.3.4, we determine the eigenvectors b^i, $i = 1, 2, 3$. It is convenient to represent these vectors as

$$N(\beta) = \|b^1, b^2, b^3\| = \begin{Vmatrix} 1 & 0 & 0 \\ 0 & \cos \beta & -\sin \beta \\ 0 & \sin \beta & \cos \beta \end{Vmatrix} \tag{4.3.10}$$

$$\beta = -\text{sign}(\sin \alpha_3)\cos^{-1}[(Q + 2 + \xi + 3\cos \alpha_2 + 2\cos^2 \alpha_3)/(2Q)]^{1/2}$$
$$Q = [4(\xi + 3)\cos^2 \alpha_3 + 6(\xi + 4)\cos \alpha_3 + 13 + \xi^2 + 4\xi]^{1/2}$$

Note that the matrix $N(\beta)$ is actually the matrix of rotation about the axis O_2X' by the angle β. Thus, the natural oscillations of the system admit the following geometrical interpretation. In one of the natural modes (corresponding to the vector b^1), the point load moves normally to the plane of the manipulator. In the other two modes corresponding to the vectors b^2 and b^3, the load oscillates in the plane of the manipulator; the directions of the motions in these modes are perpendicular to each other. These directions form the angle β with the corresponding axes Y', Z' of the reference frame $O_2X'Y'Z'$.

If the initial displacement and velocities of the load are directed along one of the vectors b^i, we have single-frequency oscillations of the ith mode. According to Equation 4.3.6, any motion of the system can be represented as a superposition of these oscillation modes.

4.3.3. ANALYSIS OF RESULTS

Let us present some computational results. For the computations, the numerical parameters describing geometrical and dynamical properties of the manipulator and the load have been taken as follows:

1) $\quad \xi = \dfrac{E_1I_1}{E_2I_2} = \dfrac{3}{2}, \zeta = \dfrac{C_1}{E_2I_2} = \dfrac{9}{8}$

$\quad l_x = l_y = l_z = 0, v_{ij} = \delta_{ij}, i,j = 1, 2, 3$

$\quad J_1 = J_2 = J_3 = 0$

2) $\quad \xi = \dfrac{E_1I_1}{E_2I_2} = \dfrac{3}{2}, \zeta = \dfrac{C_1}{E_2I_2} = \dfrac{9}{8}, k = \dfrac{C_2}{E_2I_2} = \dfrac{3}{4}$

$\quad l_x = l_y = 0, l_z/l = 1/2, v_{ij} = \delta_{ij}, i,j = 1, 2, 3$

$\quad \dfrac{J_1}{ml^2} = \dfrac{J_2}{ml^2} = \dfrac{J_3}{ml^2} = \dfrac{1}{50}$

3) $\quad \xi = 1.5, \zeta = 9/8, k = 3/4$

$\quad l_x = l_y = 0, l_z/l = 1/2, v_{ij} = \delta_{ij}, i = 1, 2, 3$

$\quad \dfrac{J_1}{ml^2} = \dfrac{3}{4}, \dfrac{J_2}{ml^2} = \dfrac{J_3}{ml^2} = \dfrac{1}{50}$ $\hfill (4.3.11)$

In case 1, the payload is a mass point placed at the end of link O_1O_2. In this case, the formulas for eigenfrequencies and shapes of natural oscillations are presented in explicit form, Equations 4.3.9 and 4.3.10.

Case 2 corresponds to the load with equal principal moments of inertia. The center of mass lies on the undeformed link O_1O_2, and is distanced from the point O_2 by half of the length of this link.

Case 3 implies that the manipulator handles a perfectly rigid body with two equal inertia moments (a dynamically symmetric body). The axis of dynamical symmetry of the load lies in the plane of the manipulator OO_1O_2 and is normal to the link O_1O_2. The distance from the point O_2 to the axis of dynamical symmetry of the load is the same as in case 2.

In all the three cases, $E_1I_1/(E_2I_2) = 1.5$, i.e., the first link is assumed to be 1.5 times as stiff as the second one. This corresponds to the strength requirements. The Poisson coefficient is $\nu = 1/3$. In cases 2 and 3, unlike 1, the frequencies have been calculated numerically by standard algorithms.

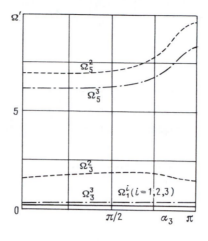

FIGURE 4.3.1.

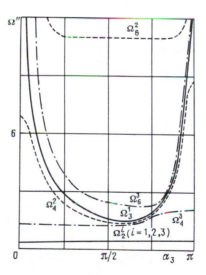

FIGURE 4.3.2.

The results of the calculations are presented in Figures 4.3.1 to 4.3.5 in dimensionless variables. Here, the abscissae axes correspond to the angle α_3 between the links OO_1 and O_1O_2. This angle can change within the interval $[-\pi, \pi]$, but, due to the symmetry, we may confine ourselves to the interval $[0, \pi]$. The ordinate axes in Figures 4.3.1 to 4.3.4 correspond to the dimensionless natural frequencies. The frequencies Ω_i^j are numbered so that the superscript $j = 1, 2, 3$ indicates the number of the case from Equations 4.3.11, while the subscript i shows serial numbers of frequencies ($\Omega_1^j < \Omega_2^j < \ldots$). Here, $i = 1, 2, 3$ for $j = 1$; $i = 1, 2, \ldots, 6$ for $j = 2, 3$.

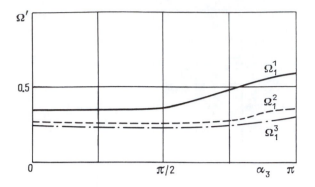

FIGURE 4.3.3.

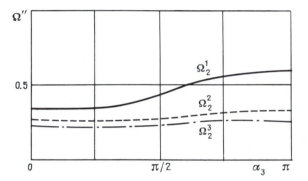

FIGURE 4.3.4.

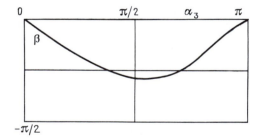

FIGURE 4.3.5.

Note that the low frequencies Ω_1^j and Ω_2^j, $j = 1, 2, 3$, are close to each other. They practically coincide in Figures 4.3.1 and 4.3.2, and are shown there by the same solid line. These frequencies are given on another scale in Figures 4.3.3 and 4.3.4. The dimensionless frequencies (Ω) are related to dimensionless ones (ω) by

$$\omega = \left(\frac{3E_1I_1}{ml^3}\right)^{1/2}\Omega$$

For all three cases in Equations 4.3.11, the oscillations are separated into two independent groups: the oscillations normal to the plane of the manipulator OO_1O_2 and the oscillations in this plane. The number of frequencies is equal to the number of degrees of freedom of the load. In case 1 the load is regarded as a point mass, and hence, the number of frequencies is 3; the first group includes one frequency Ω_1^1, while the second comprises two frequencies, Ω_2^1 and Ω_3^1. In cases 2 and 3 the load is considered as a rigid body and the number of frequencies is 6: three frequencies (Ω_1^j, Ω_3^j, Ω_5^j) belong to the first group and three frequencies (Ω_2^j, Ω_4^j, Ω_6^j) are included in the second group, $j = 2, 3$. Figures 4.3.1 to 4.3.4 show the natural frequencies Ω_i^j against the angle α_3 for the examined cases 1 to 3. Solid, dashed, and dot-and-dash lines correspond to the first ($j = 1$), second ($j = 2$), and third ($j = 3$) cases, respectively.

For cases 2 and 3 the frequencies of oscillations of the first group are given by Figures 4.3.1 and 4.3.3, while Figures 4.3.2 and 4.3.4 correspond to oscillations of the second group. Three frequencies, Ω_1^j, Ω_2^j, and Ω_4^j, $j = 2, 3$, correspond to translational motions of the load (a rigid body) and are close to the respective frequencies for case 1 (see Figures 4.3.1 to 4.3.4). Three other frequencies, Ω_i^j, $i = 3, 5, 6$; $j = 2, 3$, are due to rotations of the load. For both cases 2 and 3, the first group of frequencies includes one "translational" and two "rotational" frequencies, while the second group comprises two "translational" and one "rotational" frequencies.

The calculations show that low (basic) frequencies are close to each other in all three cases (see Figures 4.3.1 to 4.3.4). This means that the lowest frequencies are mostly due to translational motions of the load. Taking this into account, we can often consider the load as a mass point. This simplification does not cause significant errors when analyzing the dynamics of transport operations of the manipulator.

The oscillations with the lowest frequencies, Ω_1^j, $j = 1, 2, 3$, occur in the direction normal to the plane OO_1O_2. The minimum of $\Omega_1^j(\alpha_3)$, $j = 1, 2, 3$, over the angle α_3, is attained for the extended manipulator, i.e., when $\alpha_3 = 0$. As α_3 increases, the frequencies $\Omega_1^j(\alpha_3)$, $j = 1, 2, 3$ increase monotonically and assume their maximum values when $\alpha_3 = \pi$; the maximal value $\Omega_1^j(\pi)$ is approximately twice as large as $\Omega_1^j(0)$.

The oscillations with frequencies $\Omega_2^j(\alpha_3)$, $j = 1, 2, 3$ occur in the plane of the manipulator OO_1O_2. The behavior of the functions $\Omega_2^j(\alpha_3)$ is similar to that of the functions $\Omega_1^j(\alpha_3)$. The calculations show that for $j = 1, 2$, the equality $\Omega_1^j = \Omega_2^j$ takes place for the extended ($\alpha_3 = 0$) and folded ($\alpha_3 = \pi$) configurations of the manipulator. This equality is due to the symmetry of cross sections (circular) of the manipulator links and dynamical symmetry of the load (its inertia ellipsoid is a ball). Consider now the frequencies,

Ω_3^1, Ω_4^2, and Ω_4^3, which are also related to translational motions of the load. Note that $\Omega_3^1(\alpha_3) \to \infty$ as $\alpha_3 \to 0$ and $\alpha_3 \to \pi$. This is explained by the "loss" of a degree of freedom when the manipulator is completely extended or folded. The calculations show that the fourth frequency, Ω_4^2 and Ω_4^3, depends slightly on the inertia moment of the load about the axis normal to the plane of the manipulator (see Figure 4.3.2). The frequencies Ω_4^2 and Ω_4^3 tend to Ω_3^1, while Ω_6^j, $j = 2$, 3 tend to infinity as $J_1 \to 0$.

The remaining three frequencies, Ω_3^j, Ω_5^j, and Ω_6^j, $j = 1$, 2, 3 mostly depend on the inertia moments of the load. These frequencies are several times as large as the low frequencies Ω_1^j and Ω_2^j, $j = 1$, 2, 3.

Figure 4.3.5 presents the angle β (see Equation 4.3.10), determining the directions of the natural oscillations, against the angle α_3.

4.4. EQUATIONS OF MOTION FOR ELASTIC MANIPULATOR

Henceforth, throughout this chapter, we consider the case in which linear dimensions of a load carried by the manipulator are small compared to the length l of the manipulator link. Therefore, we will treat the load as a mass point. In Section 4.4, starting from the general relationships of Section 4.2, we write down the equations of motion and show that the general asymptotic method of Chapter 3 is applicable to their solution. Then, in Sections 4.5 to 4.7, we solve Problems 3.1 to 3.3 for the elastic manipulator.

The kinetic energy of the inertia-free manipulator with a point load is given by

$$T = \frac{1}{2} m\dot{\mathbf{r}}^2, \ \mathbf{r} = \overrightarrow{OO_2} \tag{4.4.1}$$

By $\mathbf{r}$ in Equation 4.4.1, we denote the radius-vector of the load (the point O_2) with respect to the origin of the inertial reference frame $OXYZ$ (see Section 4.2).

The potential energy of elastic deformation of the manipulator with a point mass at its end is given by (see Equations 4.2.24):

$$\Pi = \frac{1}{2} (C_0 \mathbf{u}_2, \mathbf{u}_2) \tag{4.4.2}$$

Here, C_0 is the stiffness matrix determined by Equations 4.2.24; $\mathbf{u}_2 = \|x_2, y_2, z_2\|$ is a vector of elastic displacement of the load represented by its components in the reference frame $O_2X'Y'Z'$. Let us recall that the reference frame $O_2X'Y'Z'$ is connected with the second link O_1O_2 of the undeformed manipulator. The axis O_2Z' is directed along the link O_1O_2, the axis O_2X' is parallel to the axis $\mathbf{e}_3$ of the elbow joint, and the axis O_2Y' complements the system to a right-hand orthogonal frame.

Let us rewrite the potential energy (Equation 4.4.2) as a function of the vector $\mathbf{u} = \|x,y,z\|$; its components are projections of the elastic displacement of the load onto the axes of the inertial reference frame $OXYZ$. The vectors $\mathbf{u}_2$ and $\mathbf{u}$ are related by

$$\mathbf{u}_2 = V\mathbf{u} \tag{4.4.3}$$

$$
V = V(\alpha)
$$
$$
= \left\|
\begin{array}{ccc}
\cos \alpha_1 & \sin \alpha_1 & 0 \\
-\sin \alpha_1 \cos(\alpha_2 + \alpha_3) & \cos \alpha_1 \cos(\alpha_2 + \alpha_3) & \sin(\alpha_2 + \alpha_3) \\
\sin \alpha_1 \sin(\alpha_2 + \alpha_3) & -\cos \alpha_1 \sin(\alpha_2 + \alpha_3) & \cos(\alpha_2 + \alpha_3)
\end{array}
\right\|
$$

Here, $\alpha = \|\alpha_1, \alpha_2, \alpha_3\|$, $V(\alpha)$ is the matrix of direction cosines of the axes of the reference frame $OXYZ$ in the frame $O_2X'Y'Z'$. Substituting $\mathbf{u}_2$ from Equation 4.4.3 into Equation 4.4.2 we obtain

$$\Pi = \frac{1}{2}(C\mathbf{u}, \mathbf{u}), \quad C = V^{\mathrm{T}} C_0 V \tag{4.4.4}$$

The kinematics of the rigid model of the manipulator is determined by the relationships

$$
r_x^0 = l \sin \alpha_1 [\sin \alpha_2 + \sin(\alpha_2 + \alpha_3)]
$$
$$
r_y^0 = -l \cos \alpha_1 [\sin \alpha_2 + \sin(\alpha_2 + \alpha_3)]
$$
$$
r_z^0 = l[\cos \alpha_2 + \cos(\alpha_2 + \alpha_3)] \tag{4.4.5}
$$

where r_x^0, r_y^0 and r_z^0 are components of the radius-vector $\mathbf{r}^0$ of the point O_2 in the reference frame $OXYZ$ when the manipulator is undeformed.

Having solved the set of Equations 4.4.5 for the angles $\alpha_i (i = 1, 2, 3)$ we obtain

$$
\alpha_1 = -\tan^{-1}(r_x^0/r_y^0), \quad r_y^0 \le 0
$$
$$
\alpha_1 = \pi - \tan^{-1}(r_x^0/r_y^0), \quad r_y^0 > 0
$$
$$
\alpha_2 = \pi + \tan^{-1}\{[(r_x^0)^2 + (r_y^0)^2]^{1/2}/r_z^0\} - \alpha_3/2, \quad r_z^0 \le 0
$$
$$
\alpha_2 = \tan^{-1}\{[(r_x^0)^2 + (r_y^0)^2]^{1/2}/r_z^0\} - \alpha_3/2, \quad r_z^0 > 0
$$
$$
\alpha_3 = \pm 2\cos^{-1}\{[(r_x^0)^2 + (r_y^0)^2 + (r_z^0)^2]^{1/2}/(2l)\} \tag{4.4.6}
$$

Equations 4.4.6 determine a configuration (joint angles) of the rigid manipulator for a given position of the load. The nonuniqueness in the expression for α_3 is due to the fact that any position of load inside the operational volume $(OO_2 < 2l)$ corresponds to two possible configurations of the manipulator.

For the elastic manipulator, we have the kinematic relationship:

$$\mathbf{r} = \mathbf{r}^0(\alpha) + \mathbf{u} \tag{4.4.7}$$

The virtual work of control torques is given by

$$\delta A = (M, \delta\alpha), \quad M = \|M_1, M_2, M_3\|, \quad \alpha = \|\alpha_1, \alpha_2, \alpha_3\| \tag{4.4.8}$$

Let us take the components, x, y, z of the vector $\mathbf{u}$ of elastic displacements and the joint angles α_1, α_2, α_3 as generalized coordinates. Then the Lagrange equations for the mechanical system determined by Equations 4.4.1, 4.4.4, 4.4.7, and 4.4.8 take the form

$$m\ddot{\mathbf{r}} + C\mathbf{u} = 0, \mathbf{r} = \mathbf{r}^0(\alpha) + \mathbf{u} \tag{4.4.9}$$

$$M = -B^{\mathrm{T}}C\mathbf{u} + \frac{1}{2}\frac{\partial(C(\alpha)\mathbf{u}, \mathbf{u})}{\partial\alpha}$$

$$B = \|b_{ij}\| = \left\|\frac{\partial r_i^0(\alpha)}{\partial\alpha_j}\right\|, i, j = 1, 2, 3$$

$$r_1^0 = r_x^0, r_2^0 = r_y^0, r_3^0 = r_z^0 \tag{4.4.10}$$

Degenerate Lagrange's Equations 4.4.10 corresponding to the components of the vector $\alpha = \|\alpha_1, \alpha_2, \alpha_3\|$ are the equilibrium conditions for an inertia-free manipulator acted upon by control torques and elastic force due to deformation of the manipulator.

Using kinematic Equation 4.4.7 we can rewrite the system of Equations 4.4.9 in one of the forms:

$$m\ddot{\mathbf{r}} + C(\alpha)(\mathbf{r} - \mathbf{r}^0(\alpha)) = 0 \tag{4.4.11}$$

$$m\ddot{\mathbf{u}} + C(\alpha)\mathbf{u} = -m\ddot{\mathbf{r}}^0(\alpha) \tag{4.4.12}$$

Let us analyze relative magnitudes of terms in Equations 4.4.9 to 4.4.12. The examined mechanical system has the following characteristic scales for length and time: l, the length of the manipulator link; u, the characteristic magnitude of elastic displacements $(x, y, z \sim u)$; T_0, time of an operation performed by the manipulator; T_1, the period of the lowest mode of elastic oscillations.

If the load travels by a distance comparable with the length of the link ($\sim l$) in the time T_0, then, as shown in Section 4.2 (see Equations 4.2.1 and 4.2.2), we have

$$u/l \sim (T_1/T_0)^2 \ll 1,\ u \sim mlT_0^{-2}c^{-1},\ T_1 \sim (m/c)^{1/2} \qquad (4.4.13)$$

Here, c is a characteristic stiffness of the manipulator structure. The elements of the matrix C in Equation 4.4.4 are quantities of the order of c. Introducing the small parameter $\epsilon = T_1/T_0$ we can rewrite Equations 4.4.13 as

$$u \sim \epsilon^2 l,\ c \sim \epsilon^{-2}mT_0^{-2},\ cu \sim mlT_0^{-2}$$

These relationships imply that the matrix C and the vector u of elastic displacement of the load can be expressed as

$$C = \epsilon^{-2}C',\ \mathbf{u} = \epsilon^2\mathbf{u}',\ \epsilon \ll 1$$
$$C' = O(1),\ \mathbf{u}' = O(1) \qquad (4.4.14)$$

In conclusion of this section, we enumerate basic mechanical properties of the elastic manipulator with a mass point (a load) at its end.

1.   The manipulator is a holonomic, scleronomous mechanical system with six degrees of freedom. The position of all points of the system is uniquely determined by the joint angles $\alpha = \|\alpha_1, \alpha_2, \alpha_3\|$ and components of the vector of elastic displacement $\mathbf{u} = \|x, y, z\|$. For given α and $\mathbf{u}$, the configuration of the manipulator is a quasistatic (equilibrium) shape of the two-link elastic structure (see Section 4.1).
2.   The system contains the inertial part, a load, and inertia-free parts, the links.
3.   The kinematics of the elastic manipulator with a load is described by Equation 4.4.7 connecting the coordinates of the load joint angles and vector of elastic displacement of the load.
4.   The manipulator is controlled by the torques M_1, M_2, M_3 at the joints. These torques are the generalized forces corresponding to the generalized coordinates (joint angles α_1, α_2, α_3) of the rigid model.
5.   The stiffness of the links is great, while elastic displacements are small. The relations between basic parameters determining the system dynamics are given by Equations 4.4.13 and 4.4.14.
6.   As $\epsilon \to 0$, the elastic model of the manipulator turns into the rigid model, elastic elements becoming additional ideal constraints.

In Chapter 3, we have studied general mechanical systems containing inertia-free elastic parts of high stiffness. The manipulator possessing properties 1 to 6 belongs to the class of systems considered in Chapter 3. In

subsequent sections of Chapter 4 we solve some problems of dynamics of
the elastic manipulator using the asymptotic technique developed in Chapter
3. The material of Section 4.5 is based on Reference 166, and the material
of Sections 4.6 and 4.7 on Reference 167.

4.5. KINEMATIC CONTROL

4.5.1. STATEMENT OF THE PROBLEM

For the mechanical system described in the previous section, we state the
following problem.

Problem 4.1. Suppose the time history of the angles $\alpha = \alpha^0(t)$ is given.
Determine the motion of the system, i.e., the functions $\mathbf{r}(t)$, $\mathbf{u}(t)$, and find
the control torques corresponding to the specified law $\alpha = \alpha^0(t)$.

In Problem 4.1, we can specify, instead of the joint angles, the motion
of the load for the rigid model $\mathbf{r} = \mathbf{r}^0(t)$. The functions $\alpha^0(t)$ and $\mathbf{r}^0(t)$ are
related by Equations 4.4.5 and 4.4.6.

Problem 4.1 is a particular case of general Problem 3.1 considered in
Section 3.1.3. Hence, we will solve Problem 4.1 using the asymptotic approach developed in Chapter 3.

First, we integrate Equation 4.4.12 with $\alpha = \alpha^0(t)$ and determine the
vector $\mathbf{u}$ of elastic displacement of the load. Then, using Equation 4.4.10,
we find the control torques $M(t) = \|M_1(t), M_2(t), M_3(t)\|$ ensuring the given
law $\alpha = \alpha^0(t)$. Finally, we determine the motion of the load $\mathbf{r} = \mathbf{r}^0(\alpha^0(t))$
$+ \mathbf{u}(t)$ using kinematic Equation 4.4.7.

4.5.2. CONSTRUCTION OF ASYMPTOTIC SOLUTION

Below we follow the method of Section 3.2.1. We assume (see Equation
3.2.1)

$$C = \epsilon^{-2}C', \mathbf{u} = \mathbf{y} + \mathbf{z}, \mathbf{y} = \epsilon^2\mathbf{y}'$$

$$\mathbf{z} = \epsilon^2\mathbf{z}'(t'), t = \epsilon t' \tag{4.5.1}$$

The fast time t' changes by an amount of the order of unity during the
period of elastic oscillations. We determine the term $\mathbf{y}$ in Equations 4.5.1 as
follows (see Equation 3.1.28):

$$\mathbf{y}(t) = -C^{-1}(\alpha)m\ddot{\mathbf{r}}^0 \tag{4.5.2}$$

Comparing Equation 4.5.2 to Equation 4.4.12, we see that $\mathbf{y}$ is a quasistatic
(slowly changing) elastic displacement. According to Equations 4.5.1,
$\mathbf{y} \sim \epsilon^2$. Substituting Equations 4.5.1 and 4.5.2 into Equation 4.4.12, we
obtain, after some simplifications, the equation for the fast (oscillatory) component of the elastic displacement (see Equation 3.2.2):

$$m \frac{d^2 z'}{dt'^2} + C'(\alpha^0)z' = 0 \qquad (4.5.3)$$

The matrix $C'(\alpha^0(t))$ depends on the slow (in comparison with t') time t. An asymptotic solution of Equation 4.5.3 with slowly changing coefficients is constructed by the averaging method and, in the first approximation, has the form of Equations 3.2.6, 3.2.7, and 3.2.10:

$$\mathbf{z} = \sum_{i=1}^{3} \boldsymbol{\varphi}^i a_i \cos \psi_i, \quad a_i = a_{i0} \left[\frac{\omega_i(t_0)}{\omega_i(t)} \right]^{1/2}$$

$$\psi_i = \int_{t_0}^{t} \varphi_i(\tau)d\tau + \psi_{i0}, \ i = 1, 2, 3 \qquad (4.5.4)$$

Here, we have returned to the initial (without primes) variables according to Equations 4.5.1. By ω_i and $\boldsymbol{\varphi}^i$ we denote the frequencies and amplitude vectors (depending on the slow time t), and by a_{i0} and ψ_{i0} the initial amplitudes and phases.

The frequencies and amplitude vectors are determined from the solution of the eigenvalue problem (see Equation 3.2.4),

$$[C(\alpha) - \omega^2 mE]\boldsymbol{\varphi} = 0 \qquad (4.5.5)$$

where E is the unit 3×3 matrix. Since $\alpha = \alpha^0(t)$, ω_i and $\boldsymbol{\varphi}^i$ also depend on t. Natural oscillations of an elastic manipulator with a point load were studied in Section 4.3. Eigenfrequencies and amplitude vectors of these oscillators are given by Equations 4.3.9 and 4.3.10.

Using Equations 4.3.10 and 4.4.3, we can represent the solution given by Equations 4.5.4 as follows:

$$\mathbf{z} = V^{\mathrm{T}}(\alpha)N(\beta)\mathbf{f}$$

$$f_i = a_{i0} \left[\frac{\omega_i(t_0)}{\omega_i(t)} \right]^{1/2} \cos \left[\int_{t_0}^{t} \omega_i(\tau)d\tau + \psi_{i0} \right], \ i = 1, 2, 3 \qquad (4.5.6)$$

The constants a_{i0} and ψ_{i0} in Equation 4.5.6 are determined by the initial data $\mathbf{u}(t_0)$ and $\dot{\mathbf{u}}(t_0)$. From Equation 4.5.1, we have

$$\mathbf{z}(t_0) = \mathbf{u}(t_0) - \mathbf{y}(t_0), \ \dot{\mathbf{z}}(t_0) = \dot{\mathbf{u}}(t_0) \qquad (4.5.7)$$

Here, the term $\dot{\mathbf{y}}(t_0)$ is dropped, since the order of this term is less than the order of $\dot{\mathbf{z}}(t_0)$. Substituting the solution given by Equation 4.5.6 into Equations 4.5.7 and taking into account that the matrices V and N are orthogonal, we obtain

$$\mathbf{f}(t_0) = [N^T V(\mathbf{u} - \mathbf{y})_{t=t_0}, \dot{\mathbf{f}}(t_0)] = [N^T V \dot{\mathbf{u}}]_{t=t_0} \qquad (4.5.8)$$

Using Equation 4.5.6, we express the constants a_{i0} and ψ_{i0} in terms of $f_i(t_0)$ and $\dot{f}_i(t_0)$:

$$a_{i0} = \{f_i^2(t_0) + [\dot{f}_i(t_0)\omega_i^{-1}(t_0)]^2\}^{1/2}$$

$$\psi_{i0} = -\tan^{-1}\left[\frac{\dot{f}_i(t_0)}{f_i(t_0)\omega_i(t_0)}\right], f_i(t_0) \geq 0$$

$$\psi_{i0} = \pi - \tan^{-1}\left[\frac{\dot{f}_i(t_0)}{f_i(t_0)\omega_i(t_0)}\right], f_i(t_0) < 0 \qquad (4.5.9)$$

When deriving Equations 4.5.8 and 4.5.9, we have neglected time derivatives of the matrices and vectors V, N, and ω_i, depending on α. These derivatives are small in comparison with the velocity of elastic oscillations, $\dot{\mathbf{u}}$.

Thus, the motion of an elastic manipulator is determined as follows. Given $\alpha^0(t)$ (or $\mathbf{r}^0(t)$), we obtain $\mathbf{r}^0(t)$ (or $\alpha^0(t)$) using Equations 4.4.5 (or 4.4.6). Then, the quasistatic displacement $\mathbf{y}(t)$ is calculated according to Equation 4.5.2. Using the initial conditions $\mathbf{u}(t_0)$ and $\dot{\mathbf{u}}(t_0)$, we determine $\mathbf{f}(t_0)$, $\dot{\mathbf{f}}(t_0)$, a_{i0}, and ψ_{i0} from Equations 4.5.8 and 4.5.9. After that, we calculate the function $z(t)$ by means of Equation 4.5.6 and determine the total elastic displacement $\mathbf{u} = \mathbf{y} + \mathbf{z}$ and the position of the load, $\mathbf{r}(t) = \mathbf{r}^0(\alpha^0(t)) + \mathbf{u}(t)$, for any time instant. The control torques $M(t)$ that implement the prescribed law $\alpha = \alpha^0(t)$ for the joint angles are determined by Equation 4.4.10. In this equation we can omit the second (quadratic in $\mathbf{u}$) term without loss of accuracy. The force $\mathbf{\Phi}$ applied to the manipulator by the load can be determined by the following formula stemming from Equation 4.4.4:

$$\mathbf{\Phi} = \frac{\partial \Pi}{\partial \mathbf{u}} = C(\alpha^0(t))\mathbf{u}(t) \qquad (4.5.10)$$

This force is equal to the reaction force at the base of the manipulator, taken with the opposite sign.

The described procedure requires only calculations by means of explicit formulas and three quadratures in Equation 4.5.6. The functions V, C, ω_i, and N, depending on the angles α_i, are determined by Equations 4.4.3, 4.4.4, 4.3.9, and 4.3.10.

Note that the functions $\alpha_i^0(t)$ and $\mathbf{r}^0(t)$ are supposed to be sufficiently smooth. In practice, however, the programs are widely used in which the acceleration $\ddot{\mathbf{r}}^0(t)$ undergoes discontinuities (jumps). At the instant t_* of the jump of $\ddot{\mathbf{r}}^0$, the displacement $\mathbf{y}$ also jumps, according to Equation 4.5.2, and therefore cannot be regarded as a slow variable. The computational technique in a vicinity of the instant t_* is altered: the vector $\mathbf{y}$ is still determined by Equation 4.5.2, but the "constants" a_{i0} and ψ_{i0} in Equation 4.5.6 jump at the instant t_*. New values of the integration constants a_{i*} and ψ_{i*} are determined for $t > t_*$ analogously to Equations 4.5.7 through 4.5.9 from the continuity conditions for the total displacement, $\mathbf{u}$, and velocity, $\dot{\mathbf{u}}$, at the instant t_*.

4.5.3. NUMERICAL SIMULATION AND DISCUSSION

For comparison, parallel with the asymptotic solution, we constructed the solution of Problem 4.1 by direct numerical integration of Equation 4.4.12. We specified the initial data for $\mathbf{u}$ and $\dot{\mathbf{u}}$ at $t = t_0 = 0$, and the function $\mathbf{r}^0(t)$ describing the motion of the load for the manipulator with rigid links. The laws for the joint angles $\alpha_i = \alpha_i^0(t)$, i = 1, 2, 3, were calculated by using Equations 4.4.6. The integration was carried out by the Runge-Kutta method with automatic choice of step and accuracy control. After determining the elastic displacement $\mathbf{u}(t)$, we calculated the force $\boldsymbol{\Phi}(t)$ and torques $M(t)$ according to Equations 4.5.10 and 4.4.10.

Let us compare the numerical and asymptotic approaches. Unlike the asymptotic technique, the numerical one does not require determining eigen-frequencies and shapes of oscillation modes, inverse for the stiffness matrix (see Equation 4.5.2), and initial data in Equations 4.5.7 through 4.5.9. On the other hand, the asymptotic technique allows us to obtain the solution in the explicit form. To calculate the quadratures in the asymptotic solution given by Equations 4.5.6, we can take the integration step much further than the step of integrating the equations of elastic oscillations (Equation 4.4.12). In our case, the eigenfrequencies and shapes of oscillation modes, as well as the inverse for the matrix $\mathbf{C}$, are determined beforehand in the analytical form.

Now let us consider and discuss some computational results. We introduce the dimensionless variables, taking the magnitudes m, l, and $[3E_1I_1/(ml)^3]^{1/2}$ as new standards for mass, length, and time, respectively. In the dimensionless variables, we have

$$l = 1, \; m = 1, \; E_1I_1 = 1/3 \qquad (4.5.11)$$

We take the parameters ξ and ζ introduced in Equations 4.2.24 as follows:

$$\xi = E_1I_1/(E_2I_2) = 3/2, \; \zeta = C_1/(E_2I_2) = 1.1 \qquad (4.5.12)$$

We specify the motion of the load for the rigid model as $\mathbf{r}^0(t) = \boldsymbol{\rho}(t)$. Here, the trajectory $\boldsymbol{\rho}(t)$ is a broken line $K_0K_1K_2$ consisting of two segments of equal length which lie in the plane $X = -0.45$ and form the angle $120°$. The coordinates of the points K_0, K_1, and K_2 in the inertial reference frame $OXYZ$ are given by (see Figure 4.5.1):

$$K_0 = (-0.45, 0.84, 0.52), \; K_1 = (-0.45, 0.33, 0.82)$$

$$K_2 = (-0.45, 0.33, 1.42) \qquad (4.5.13)$$

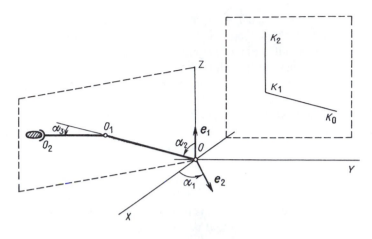

FIGURE 4.5.1.

The load moves along both segments K_0K_1 and K_1K_2 in the same way. First, it accelerates for the dimensionless time $\tau_1 = 17$, then moves with a constant velocity for $\tau_2 = 71$, and then decelerates for $\tau_3 = 17$. The constant velocity at the middle stages is equal to $|\dot{\rho}| = v\ 0.0068$. At the acceleration and deceleration stages, the velocity changes linearly from 0 to v and from v to 0, respectively. The total duration of the operation is $T = 2(\tau_1 + \tau_2 + \tau_3) = 210$.

In the course of motion, the acceleration $\ddot{\rho}$ undergoes a number of jumps. When using the asymptotic approach, it is necessary to find new initial data at the instants when $\ddot{\rho}$ is discontinuous, as described at the end of Section 4.5.2.

The results obtained by both the asymptotic and numerical methods are close to each other.

Some results are presented in Figures 4.5.2 through 4.5.4. Figure 4.5.2 shows the projections of the vector **u** of elastic displacement (against the time) onto the axes of the inertial reference frame *OXYZ* (curves 1). In this picture, we also give the projections of the quasistatic displacement **y**(*t*) (see Equation 4.5.2, onto the same axes (curves 2). We can see that the quasistatic displacement **y**(*t*) jumps at the discontinuity points of the acceleration $\ddot{\rho}$, while the total elastic displacement **u**(*t*) is a continuous smooth function. The function **u**(*t*) contains oscillations with slowly changing amplitude and frequency. These oscillations correspond mostly to low frequencies ω_1 and ω_2 from Equations 4.3.9.

Figure 4.5.3 presents the projections of the force **Φ** onto the axes of the moving reference frame $O_2X'Y'Z'$ (curves 1). In the projection of **Φ** onto the axis O_2Z', the oscillations of the higher mode (with the frequency ω_3) are considerable. Figure 4.5.4 shows the joint torques $M_i(t)$, $i = 1, 2, 3$ (curves 1). Here, high-frequency oscillations for $M_2(t)$ are significant.

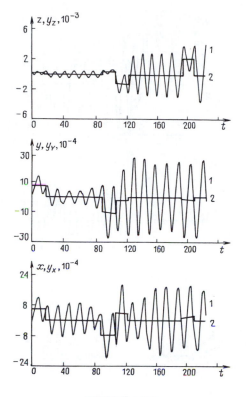

FIGURE 4.5.2.

For comparison, in Figures 4.5.3 and 4.5.4 we also show the time history of the force $\boldsymbol{\Phi}$ and the torques M_i, $i = 1, 2, 3$, for the rigid model (curves 2). These curves do not contain oscillations and have jumps at the instants when the acceleration $\ddot{\boldsymbol{\rho}}$ is discontinuous.

4.6. DYNAMIC CONTROL

In this section we consider the following problem.

Problem 4.2. Let the torques at the joints of the manipulator be specified as functions of time: $M(t) = \|M_1(t), M_2(t), M_3(t)\|$. Determine the motion of the load, $\mathbf{r}(t)$, the vector $\mathbf{u}(t)$ of its elastic displacement, the joint angles $\alpha_i(t)$, $i = 1, 2, 3$, and the force $\boldsymbol{\Phi}$ applied to the manipulator by the load.

Problem 4.2 corresponds to Problem 3.2 of determining the motion of the system with elastic parts in the case of dynamic control (see Section 3.1.3).

For compactness, we present kinematic Equations 4.4.5 and 4.4.6 in the vector form:

$$\mathbf{r}^0 = \mathbf{f}(\alpha), \quad \alpha = f^{-1}(\mathbf{r}^0) = h(\mathbf{r}^0) \tag{4.6.1}$$

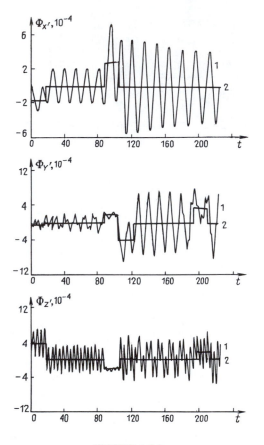

FIGURE 4.5.3.

Here, $h(\mathbf{r}^0) = f^{-1}(\mathbf{r}^0)$ is the inverse for the vector function $\mathbf{f}(\alpha)$. According to Equations 4.4.6, the function $h(\mathbf{r}^0)$ has two branches. This reflects the fact that for any position of the load, except those corresponding to the completely extended or folded manipulator, two different configurations of the manipulator are possible which differ by the sign of the angle α_3 at the elbow joint. To be definite we choose the branch corresponding to $\alpha_3 > 0$. Taking into account kinematic Equation 4.4.7 for the elastic manipulator, we can rewrite the second Equation 4.6.1 as

$$\alpha = h(\mathbf{r} - \mathbf{u}) \qquad (4.6.2)$$

Substituting the expression given by Equations 4.4.14 for the stiffness matrix C and the elastic displacement $\mathbf{u}$ into Lagrangian Equations 4.4.9 and 4.4.10, as well as into Equation 4.6.2, we obtain

$$m\ddot{\mathbf{r}} = -C'(\alpha)\mathbf{u}' \qquad (4.6.3)$$

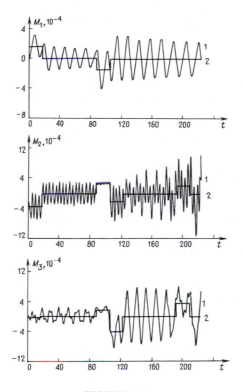

FIGURE 4.5.4.

$$M = -B^{\mathrm{T}}(\alpha)C'(\alpha)\mathbf{u}' + \frac{\epsilon^2}{2}\frac{\partial(C'(\alpha)\mathbf{u}', \mathbf{u}')}{\partial\alpha} \tag{4.6.4}$$

$$\alpha = h(\mathbf{r} - \epsilon^2\mathbf{u}') \tag{4.6.5}$$

Let us derive now the set of equations for the vector $\mathbf{r}$. To do this, we eliminate the vectors $\mathbf{u}'$ and α from Equation 4.6.3 using Equations 4.6.4 and 4.6.5. We seek $\mathbf{u}'$ as a series expansion: $\mathbf{u}' = \mathbf{u}'_0 + \epsilon^2\mathbf{u}'_1 + \ldots$. Substituting this series into Equation 4.6.4 and comparing the coefficients of like powers of the small parameter ϵ, we obtain

$$\mathbf{u}' = \mathbf{u}'_0 + \epsilon^2\mathbf{u}'_1 + O(\epsilon^4)$$

$$\mathbf{u}'_0(\alpha, t) = -[B^{\mathrm{T}}(\alpha)(C'(\alpha)]^{-1}M(t)$$

$$\mathbf{u}'_1(\alpha, t) = \frac{1}{2}[B^{\mathrm{T}}(\alpha)C'(\alpha)]^{-1}\frac{\partial(C'(\alpha)\mathbf{u}'_0, \mathbf{u}'_0)}{\partial\alpha} \tag{4.6.6}$$

Substituting $\mathbf{u}'$ from Equation 4.6.6 into Equations 4.6.3 and 4.6.5 and retaining the terms up to $O(\epsilon^2)$, we have

$$m\ddot{\mathbf{r}} = -C'(\alpha)(\mathbf{u}_0' + \epsilon^2\mathbf{u}_1') \qquad (4.6.7)$$

$$\alpha = h[\mathbf{r} - \epsilon^2\mathbf{u}_0'(\alpha, t)] \qquad (4.6.8)$$

Let us solve Equation 4.6.8 for α with the accuracy up to the terms $O(\epsilon^2)$ inclusively, taking into account the Equation 4.6.6 for $\mathbf{u}'_0$. We seek the solution as a series $\alpha = \alpha_0 + \epsilon^2\alpha_1 + \dots$. Substituting this series into Equation 4.6.8 and comparing the coefficients of like powers of ϵ, we get

$$\alpha_0 = h(\mathbf{r}), \ \alpha_1 = -\frac{\partial h(\mathbf{r})}{\partial \mathbf{r}} \mathbf{u}_0'(\alpha_0, t)$$

$$\alpha(\mathbf{r}, t) = \alpha_0 + \epsilon^2\alpha_1 + \dots = h(\mathbf{r}) - \epsilon^2 \frac{\partial h(\mathbf{r})}{\partial \mathbf{r}} \mathbf{u}_0'(\alpha_0, t) + \dots$$

$$= h(\mathbf{r} - \epsilon^2\mathbf{u}_0'(h(\mathbf{r}), t)) + O(\epsilon^4) \qquad (4.6.9)$$

Finally, replacing α in Equation 4.6.7 by the expression $\alpha(\mathbf{r}, t)$ from Equation 4.6.9 we arrive at the differential equation for the radius-vector of the load $\mathbf{r}(t)$:

$$m\ddot{\mathbf{r}} = -C'(\alpha)(\mathbf{u}_0' + \epsilon^2\mathbf{u}_1') \qquad (4.6.10)$$

We solve Problem 4.2 using the following procedure. First, we integrate Equation 4.6.10 for $\mathbf{r}$, replacing α, $\mathbf{u}_0'$ and $\mathbf{u}_1'$ by their expressions from Equations 4.6.6 and 4.6.9. The initial conditions $\mathbf{r}(t_0)$ and $\dot{\mathbf{r}}(t_0)$ are specified at the time instant $t = t_0$. As a result, we determine the motion of the load, $\mathbf{r}(t)$, as well as the law $\alpha = \alpha(t)$ of changing the joint angles. The elastic displacement $\mathbf{u}$ is expressed by the formula

$$\mathbf{u} = \epsilon^2\mathbf{u}' = \epsilon^2\mathbf{u}_0' + \epsilon^4\mathbf{u}_1' + o(\epsilon^4)$$

stemming from Equations 4.4.14 and 4.6.6. The force $\mathbf{\Phi}$ applied to the manipulator by the load is determined by Equation 4.5.10:

$$\mathbf{\Phi} = \frac{\partial \Pi}{\partial \mathbf{u}} = C(\alpha)\mathbf{u} = C'(\alpha)\mathbf{u}' \qquad (4.6.11)$$

Thus, we have described the algorithm for solving Problem 4.2. The vectors $\mathbf{r}$, α, and $\mathbf{\Phi}$ are determined with the error $O(\epsilon^4)$, while $\mathbf{u}$ is determined with the error $o(\epsilon^4)$. Note that the above method of constructing the solution concretizes the general approach to solution of Problem 3.2 expounded in Section 3.2.3 (for the case $N = n$).

The amount of calculations necessary for solving Problem 4.2 following the above procedure is almost the same as for the case of the rigid model.

We are to integrate the sixth-order system of differential equations and determine the function $\mathbf{u}(t)$ by means of explicit formulas.

Now we present some results of the numerical solution of Problem 4.2 for the elastic manipulator whose geometrical and mechanical (dimensionless) parameters are given by Equations 4.5.11 and 4.5.12. We specify the time history of the control torques, $M_i(t)$, $i = 1, 2, 3$, so that the load carried by the manipulator with rigid links moves in accordance with the law $\mathbf{r} = \boldsymbol{\rho}(t)$ described in Section 4.5.3. The corresponding trajectory of the point O_2 is the broken line $K_0 K_1 K_2$ shown in Figure 4.5.1. The coordinates of the vertices of this line are given by Equations 4.5.13. The velocity $\dot{\boldsymbol{\rho}}$ is a continuous, piecewise linear function, while the acceleration $\ddot{\boldsymbol{\rho}}$ is a discontinuous, piecewise constant function of time (see Section 4.5.3).

Assuming $\epsilon = 0$ and $\mathbf{r} = \boldsymbol{\rho}(t)$ in Equations 4.6.3 through 4.6.5, we obtain

$$M(t) = M^0(t) = B^{\mathrm{T}}[h(\boldsymbol{\rho}(t))]\ddot{\boldsymbol{\rho}}(t) \qquad (4.6.12)$$

The graphs for $M^0(t)$ specified by Equation 4.6.12 are given in Figure 4.6.1. The functions $M_i^0(t)$, $i = 1, 2, 3$ have jumps at the instants of discontinuities of the acceleration $\ddot{\boldsymbol{\rho}}$. Figure 4.6.2 shows the time history of the joint angles $\alpha_i^0(t)$, $i = 1, 2, 3$ for the manipulator with rigid links, provided the load moves according to the law $\mathbf{r} = \boldsymbol{\rho}(t)$. The functions $\alpha_i^0(t)$ are determined by

$$\alpha^0(t) = \|\alpha_1^0(t), \alpha_2^0(t), \alpha_3^0(t)\| = h(\boldsymbol{\rho}(t)) \qquad (4.6.13)$$

We have constructed the solution of Problem 4.2 by numerical integration of Equation 4.6.10 in which $\mathbf{u}'_0$, $\mathbf{u}'_1$, and α are defined by Equations 4.6.6 and 4.6.9 and the control torques are determined by Equation 4.6.12. The integration was carried out by the Runge-Kutta method with automatic choice of integration step and accuracy control.

The initial data for Equation 4.6.10 were taken as

$$\mathbf{r}(0) = \boldsymbol{\rho}(0), \ \dot{\mathbf{r}}(0) = \dot{\boldsymbol{\rho}}(0) = 0 \qquad (4.6.14)$$

If the manipulator links are rigid ($\epsilon = 0$), then the solution of Equation 4.6.10 with initial conditions given by Equations 4.6.14 is $\mathbf{r} = \mathbf{r}^0(t) = \boldsymbol{\rho}(t)$, in accordance with the choice of the control torques in Equation 4.6.12.

Figures 4.6.3 and 4.6.4 present some computational results for the elastic manipulator. Figure 4.6.3 shows time histories of projections of the vector $\Delta\mathbf{r} = \mathbf{r}(t) - \boldsymbol{\rho}(t)$ onto the axes of the inertial reference frame $OXYZ$. The vector $\Delta\mathbf{r}$ characterizes the influence of elasticity on the displacement of the load.

Note that, according to Equations 4.4.7 and 4.6.1,

$$\mathbf{u} = \mathbf{r} - \mathbf{r}^0 = \mathbf{r} - \mathbf{f}(\alpha)$$

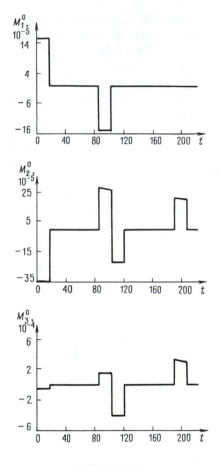

FIGURE 4.6.1.

and $\mathbf{f}(\alpha) = \boldsymbol{\rho}$ only for the manipulator with rigid links. In general, $\mathbf{f}(\alpha) \neq \boldsymbol{\rho}$ for the elastic manipulator, and therefore $\mathbf{u} \neq \mathbf{r} - \boldsymbol{\rho} = \Delta\mathbf{r}$. The function $\Delta\mathbf{r}(t)$ in Figure 4.6.3 does not contain high-frequency elastic oscillations, unlike the case of kinematic control (see Figure 4.5.2).

In Figure 4.6.4, solid lines show the functions $\Delta\alpha_i(t) = \alpha_i(t) - \alpha_i^0(t)$, $i = 1, 2, 3$, characterizing the influence of the elasticity on the joint angles. Note that the functions $\Delta\alpha_i(t)$ are continuous at the instants of jumps of the torques $M_i(t)$. The angles $\alpha_i(t)$ also have jumps at the same time instants, i.e., the configuration of the manipulator changes abruptly. This fact is explained as follows. Since the manipulator is assumed inertia-free, its current configurations are quasistatic and determined from equilibrium conditions. Therefore, when the joint torques jump, the configuration follows these jumps immediately. To obtain a more realistic description of the behavior of the manipulator, we have to take into account the inertia of the links.

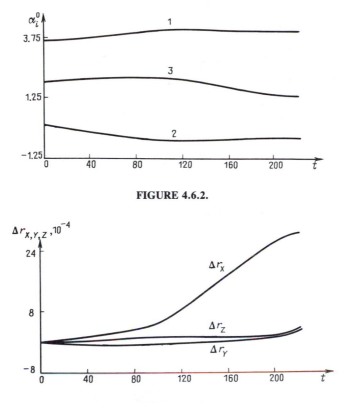

FIGURE 4.6.2.

FIGURE 4.6.3.

4.7. TRACKING THE MOTION OF THE LOAD

For a manipulator with a point load, let us consider Problem 3.3 stated in Section 3.1.3.

Problem 4.3. Suppose the desired motion of the load is specified: $\mathbf{r} = \mathbf{r}(t)$. Determine the control law $M(t)$, ensuring the given motion of the load, the elastic displacement $\mathbf{u}(t)$, joint angles $\alpha(t)$, and the force $\mathbf{\Phi}(t)$ applied to the manipulator by the load.

Let us solve Equation 4.6.3 for $\mathbf{u}'$,

$$\mathbf{u}' = -m[C'(\alpha)]^{-1}\ddot{\mathbf{r}} \tag{4.7.1}$$

and substitute $\mathbf{u}'$ from Equation 4.7.1 into Equation 4.6.5. We obtain

$$\alpha = h(\mathbf{r} + \epsilon^2 m[C'(\alpha)]^{-1}\ddot{\mathbf{r}} \tag{4.7.2}$$

To solve Equation 4.7.2 for α, we use the method applied to Equation 4.6.8. Within the accuracy $O(\epsilon^4)$, the solution can be presented as

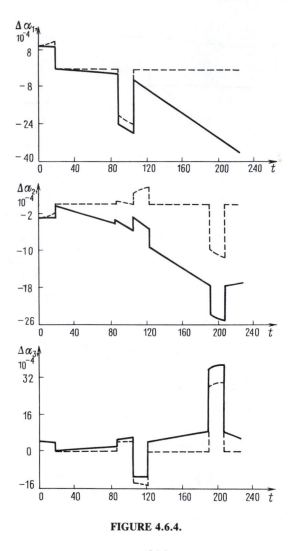

FIGURE 4.6.4.

$$\alpha(t) = h(\mathbf{r}) + \epsilon^2 m \frac{\partial h(\mathbf{r})}{\partial \mathbf{r}} [C'(h(\mathbf{r}))]^{-1}\ddot{\mathbf{r}}$$

$$= h(\mathbf{r} + \epsilon^2 m[C'(h(\mathbf{r}))]^{-1}\ddot{\mathbf{r}}), \quad \mathbf{r} = \mathbf{r}(t)$$

(4.7.3)

Substituting $\alpha(t)$ from Equation 4.7.3 into Equation 4.7.1, we determine the vector $\mathbf{u}'(t)$, and then, using Equation 4.6.4 we find the control torques ensuring the motion $\mathbf{r} = r(t)$:

$$M(t) = -B^{\mathrm{T}}C'(\alpha)\mathbf{u}' + \frac{\epsilon^2}{2} \frac{\partial(C'(\alpha)\mathbf{u}', \mathbf{u}')}{\partial \alpha}$$

(4.7.4)

Here, $\mathbf{u}' = \mathbf{u}'(t)$ is determined by Equations 4.7.1 and 4.7.3.

According to Equations 4.7.1 and 4.4.14, the vector of elastic displacement **u** is

$$\mathbf{u}(t) = \epsilon^2 \mathbf{u}' = -\epsilon^2 m[C'(\alpha)]^{-1}\ddot{\mathbf{r}} \qquad (4.7.5)$$

The force $\mathbf{\Phi}(t)$ applied to the manipulator by the load is determined by Equation 4.6.11:

$$\mathbf{\Phi}(t) = C(\alpha)\mathbf{u}(t) = C'(\alpha)\mathbf{u}'(t) \qquad (4.7.6)$$

Equations 4.7.3 through 4.7.6 give the solution of Problem 4.3. This solution is reduced to calculations according to explicit formulas and does not require integrating differential equations. The solving procedure described here corresponds to the general approach to Problem 3.3 (see Section 3.2.4).

Let us consider some computational results for the elastic manipulator whose geometrical and mechanical parameters are given by Equations 4.5.11 and 4.5.12. We specify the motion of the load by the equation $\mathbf{r}(t) = \mathbf{\rho}(t)$. The function $\mathbf{\rho}(t)$ is described in Section 4.5.3.

For the rigid model of the manipulator ($\epsilon = 0$), the control torques $M_i^0(t)$ and the angles $\alpha_i^0(t)$, $i = 1, 2, 3$, corresponding to the motion $\mathbf{r} = \mathbf{\rho}(t)$, are determined by Equations 4.6.12 and 4.6.13, respectively. The graphs of the functions $M_i^0(t)$ and $\alpha_i^0(t)$, $i = 1, 2, 3$, are shown in Figures 4.6.1 and 4.6.2. The functions $\Delta\alpha_i(t) = \alpha_i(t) - \alpha_i^0(t)$, $i = 1, 2, 3$, characterizing the influence of the elastic compliance on the joint angles of the manipulator, are given by dashed lines in Figure 4.6.4. The functions $\Delta\alpha_i(t)$ are discontinuous, as in the case of dynamic control.

In Figure 4.7.1, we present the time history of the additional torques $\Delta M_i = M_i(t) - M_i^0(t)$, $i = 1, 2, 3$, necessary for tracking the prescribed motion of the load $\mathbf{r} = \mathbf{\rho}(t)$ by the elastic manipulator. These functions jump simultaneously with the respective torques $M = M_i^0(t)$ presented in Figure 4.6.1.

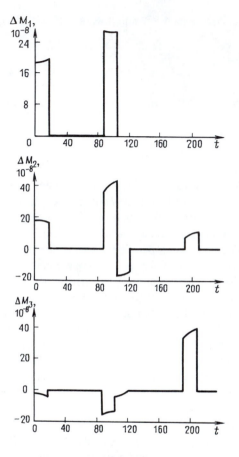

FIGURE 4.7.1.

Chapter 5

EXPERIMENTAL STUDY OF ELASTIC PROPERTIES OF MANIPULATION ROBOTS

5.1. ELASTIC COMPLIANCE OF ROBOTS

As we have already mentioned, in the majority of mechanical models, manipulation robots are regarded as systems of rigid bodies connected by joints. However, structural elements (links, joints, gear trains, etc.) of real robots possess an elastic compliance that causes static displacements of the structure under the action of gravity and, moreover, gives rise to oscillations of a manipulator when in motion. Amplitudes of such oscillations may be great, especially if frequencies of external disturbances or the frequency of feeding control signals are close to eigenfrequencies of elastic oscillations. This reduces the positioning accuracy of a manipulator and worsens its behavior when accomplishing technological operations.

Two ways are possible for mitigating unwanted effect of the elastic compliance of robots. First, links, joints, and gears can be made stiffer. This is just the choice the designers of industrial robots mostly make. However, such a way leads to excessive metal expenditure, the mass of the structure increases, and, as a consequence, energy consumption grows while speed reduces. The requirements of saving metal and increasing stiffness contradict each other.

Let us consider, for example, the parameters of the arm of a typical cylindrical-coordinate industrial robot. The arm is made of a steel tube with inner and outer diameters equal to $d = 60$ mm and $D = 70$ mm, respectively. The length of the arm is 0.8 m and the mass is 6.3 kg. The angular acceleration of the arm on its start stage is $\epsilon = 3.3$ s^{-2}.

The calculations show that if the manipulator carries a load of 10 kg, the maximal elastic displacement of the gripper on the start stage makes up about 0.1 mm, while the fatigue ratio is $n_r \approx 39$. It is many times as much as the strength of materials requires. For the majority of structures, the fatigue ratio is of the order of unity (typical value is $n_r \approx 2$). To ensure the value $n_r \approx 2$ for the fatigue ratio, it is sufficient to take the inner and outer diameters of the arm equal to $d = 22.5$ mm and $D = 25$ mm, respectively, the mass of the arm reducing tenfold. However, the structure stiffness will also drop significantly, and the maximal elastic displacement will increase to 7 mm approximately, thus exceeding the admissible positioning error many times.

The second way is to develop special control modes, suppressing elastic oscillations or reducing their amplitudes to an acceptable level. To design such control modes, we have to use a more accurate mechanical model of a robot, taking into account its flexibility.

To assess the influence of the elastic compliance of a robot on its dynamical characteristics and to choose an adequate mechanical model, we need to carry out experimental investigations of elastic properties of the robot. The total elastic compliance of a robot structure is contributed to by the flexibility of its links, joints, junctions, and gear trains. Since the length of manipulator links is much greater than linear dimensions of joints, we can regard the elasticity of joints as lumped. On the other hand, the elasticity of links is distributed, and here the models are based on partial differential equations. Relative contribution of links, joints, junctions, gears, etc., to the total compliance of a manipulator is important for the choice of an adequate mathematical model of a robot. The experiments described in Chapter 5 show that, for the majority of industrial robots (their load-carrying capacity and lengths of links being within 1 to 25 kg and 0.5 to 2 m, respectively), the compliance is mostly conditioned by the joints. The elasticity of links is not very essential, in view of their comparatively small lengths and high stiffnesses. At the same time, for large space manipulators, the elasticity of links makes the main contribution. Perhaps, the elastic compliance of links will be essential for prospective industrial robots with light links.

In subsequent sections of this chapter, we discuss methods and results of the experimental study of elastic properties for some types of industrial manipulation robots.

5.2. TECHNIQUE OF EXPERIMENTS

The experimental investigations consist of static and dynamic testing.

In static experiments, static forces varying within admissible bounds are applied to the gripper or other parts of a robot, while the latter remains in a definite configuration. Using a displacement sensor (an indicator head), we measure vertical and horizontal elastic displacements at certain points of the robot. To estimate the contribution of each individual link, joint, or junction to the total elastic compliance, we fix the unloaded parts by means of additional rigid supports. We choose the modes of loading, the ways of fastening the supports and placing the measurement heads, in accordance with the manipulator design, as described below. The displacement measurements are carried out several times for each loading mode and the results are averaged. As a rule, the averaged elastic displacements linearly depend on the force applied. This enables us to use linear theory of elastic bending of beams. To determine the shape of a deformed link, we make measurements at several points of the link.

To calculate stiffness parameters of a robot, we use equations of elastic systems equilibrium. In these calculations, links are considered as rigid or elastic bars (if the compliance of the links is essential). The elastic torque at a joint is determined by the formula $M = cu$, where u is the elastic angular displacement at the joint, while c is its stiffness. Knowing the applied forces and measuring the elastic displacements, we can evaluate the stiffness, e.g., express c through M and u.

Let us describe the general procedure for determining stiffness coefficients for a system with a finite number of degrees of freedom. Denote by $\alpha = (\alpha_1, \ldots, \alpha_n)$ the generalized coordinates (for example, joint angles) describing relative displacements in kinematic pairs, and by $u = (u_1, \ldots, u_m)$ the elastic displacements (for example, the components of the gripper elastic displacement). In the course of the experiment, we fix joint angles and apply specified forces $\mathbf{F}_j$, $j = 1, \ldots, N$, to a manipulator at N points; their radii-vectors are denoted by $\mathbf{r}_j$, $j = 1, \ldots, N$. After equilibrium has set in, we measure the elastic displacements u.

The vectors $\mathbf{r}_j$ can be expressed through α and u, using kinematic relations:

$$\mathbf{r}_j = \mathbf{f}_j(\alpha, u), j = 1, \ldots, N \qquad (5.2.1)$$

Provided the stiffness of the manipulator is high and Equations 4.2.1 and 4.2.2 hold, the equilibrium equations of the system subjected to the external forces $\mathbf{F}_j$ and elastic forces can be linearized in u and presented as follows:

$$\sum_{j=1}^{N} \mathbf{F}_j \frac{\partial \mathbf{f}_j(\alpha, 0)}{\partial u_i} = \sum_{j=1}^{m} c_{ij} u_j, i = 1, \ldots, m \qquad (5.2.2)$$

Here, c_{ij} are elements of the matrix C of the elastic potential energy (stiffness matrix).

Note that, apart from the forces $\mathbf{F}_j$, a manipulator is subjected to gravity forces, which also cause elastic displacements. Since the static Equations 5.2.2 are linear, the displacements due to the forces $\mathbf{F}_j$ and to the weight of the manipulator are summed. Therefore, in what follows we ignore the gravity forces in static equations and take into account only the displacements caused by the forces $\mathbf{F}_j$. Accordingly, in the experiments described below we measure only the displacements due to the loading forces.

Among the parameters included in Equations 5.2.2, only c_{ij}, $i, j = 1, \ldots, m$, are unknown. Since the matrix C is symmetric, there are $m(m + 1)/2$ unknown elements c_{ij}. Having carried out a sufficiently large number of measurements of elastic displacements under different loading modes, we obtain a sufficient number of Equations 5.2.2 for c_{ij}.

In practice, it is often convenient to directly measure stiffnesses of individual joints, instead of elements of the matrix C. The experiments (the choice of configurations of a manipulator, the modes of supporting and loading, etc.) should be planned so as to determine the desired coefficients with a sufficient accuracy and, at the same time, to avoid solving high-order systems of equations. For each stiffness coefficient c, we carry out a series of n independent experiments and find a series of values c_k for c. Then, the mean value, $\langle c \rangle$, and the standard deviation, σ, are determined by

$$\langle c \rangle = \frac{1}{n} \sum_{k=1}^{n} c_k, \sigma = \left[\frac{1}{n-1} \sum_{k=1}^{n} (c_k - \langle c \rangle)^2 \right]^{1/2} \qquad (5.2.3)$$

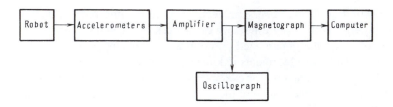

FIGURE 5.2.1.

Thus, we can obtain all elements of the desired stiffness matrix C and estimate the accuracy of their determination.

Since the kinematic structure and materials of a robot are known, we can find the inertia parameters of the robot: coordinates of mass centers of the links, masses of the links, and their inertia moments. Then we can calculate the kinetic energy for the manipulator, using the technique of Chapter 2. Having known the kinetic and potential energy matrices A and C, we can find eigenfrequencies and shapes of the robot elastic oscillations, following Section 4.3.

Dynamic experiments are aimed at investigating oscillations of a robot during its motion. For this purpose, accelerometers are set at different points of a robot. As a rule, three accelerometers are placed at each point for measuring accelerations in three orthogonal directions. Output signals of the accelerometers are recorded on a magnetic tape for subsequent processing and visualization by means of *XY*-recorders, mirror-galvanometer oscillographs, or computers.

We conducted the dynamical experiments for robots like Ritm, Universal-5 and RPM-25 (see Section 1.1.3). The motion of each degree of freedom of these robots included acceleration, constant speed, and deceleration stages. On each stage, elastic oscillations are excited in the robot structures and result in vibrations of the grippers. We conducted the measurements from the start of the motion till its termination. At the beginning and the end of the motion, oscillations were more intensive than at the middle stage. The time history of the oscillations was rather complex, but showed a good repeatability. This enabled us to get sufficient experimental data by making measurements on a single run of the robot.

Dynamic experiments were carried out within the frequency range from 1 to 600 Hz, the amplitudes of elastic oscillations varying from 0.01 to 3 mm. The measurement data were fed into the computer from the magnetograph by means of an analog-to-digital converter. The computer processed the experimental data and carried out the spectral analysis of the oscillations. As a result, we determined the spectrum of elastic oscillations, obtained the eigenfrequencies, and found out how they depend on the robot configurations. The block diagram of the installation for dynamic experiments is shown in Figure 5.2.1.

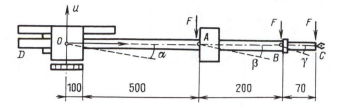

FIGURE 5.3.1.

5.3. ELASTIC COMPLIANCE OF A PNEUMATIC MANIPULATOR

We consider here static experiments for the industrial robot Ritm with pneumatic drives (see Section 1.1.3). A hand of the robot is shown in Figure 5.3.1. In the course of static experiments, we apply a force F to different characteristic points of the arm, in particular to the junctions A, B, and C. We assume that:

1. The basic link DA driven by an air piston is a cantilever beam clamped at the junction O with the stiffness c_O.
2. The links AB and BC (a hand and a gripper) can be regarded as rigid, in view of their comparatively small linear dimensions.
3. The links AB and BC are fastened by the junctions A and B with stiffnesses c_A and c_B, respectively.

If the force F is applied to the point A, the torque,

$$\alpha c_O = F(L - x) \tag{5.3.1}$$

appears at the junction O. Here, α is the angle of rotation of the link DA in the vertical plane, L is the length of the link DA, $L - x$ is the length of the extended part, OA, of the link DA, $x = |DO|$.

Having fixed the link DA at the junction O rigidly by means of a special support, we can estimate the distributed elasticity of the link, applying the force F to the point A.

Denote by EI the bending stiffness of the link DA, and by $u(\xi)$ its elastic displacement. The equilibrium conditions are given by

$$EIu'''(\xi) = -F, \, u(0) = 0, \, u'(0) = 0$$
$$u''(L - x) = 0, \, \xi \in [0, L - x] \tag{5.3.2}$$

Here, primes mean differentiation with respect to the coordinate ξ. Integrating the equation with boundary conditions (Equation 5.3.2), we obtain

$$u(\xi) = -\frac{F}{6EI}\xi^3 + \frac{F(L-x)}{2EI}\xi^2 \qquad (5.3.3)$$

Substituting $\xi = L - x$ into Equation 5.3.3, we find the displacement of the point A:

$$u_A = \frac{F}{3EI}(L-x)^3 \qquad (5.3.4)$$

Having measured u_A, we can calculate the bending stiffness of the link DA:

$$EI = \frac{F(L-x)^3}{3u_A} \qquad (5.3.5)$$

The total displacement of the point A, due to both the lumped compliance of the junction O and the distributed elasticity of the link DA, can be represented as

$$U_A = u_A + \alpha(L-x) \qquad (5.3.6)$$

where u_A is given by Equation 5.3.4. Substituting α from Equation 5.3.1 into Equation 5.3.6, we find:

$$U_A = u_A + F(L-x)^2/c_O \qquad (5.3.7)$$

From Equation 5.3.7 we can get the formula for the stiffness c_O of the junction O:

$$c_O = F(L-x)^2/(U_A - u_A) \qquad (5.3.8)$$

Since the links AB and BC are regarded as rigid, the stiffnesses of the junctions A and B can be evaluated as follows. Let us fix the first link (DA) rigidly by means of a special support and apply the force F to the end of the second link (AB). Then the torque appearing at the junction A is equal to

$$c_A\beta = Fl_1 \qquad (5.3.9)$$

where $\beta = u_B/l_1$ is the angle of rotation of the link AB in the vertical plane, l_1 is the length of the link AB, and u_B is the displacement of the point B. Substituting $\beta = u_B/l_1$ into Equation 5.3.9, we obtain:

$$c_A = Fl_1^2/u_B \qquad (5.3.10)$$

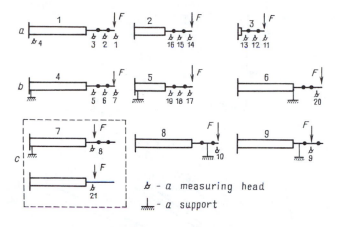

FIGURE 5.3.2.

Analogously, fixing the end of the second link of the manipulator arm and applying the force F to the gripper, we can evaluate the stiffness c_β of the junction B as follows:

$$c_\beta = Fl_2^2/u_c \qquad (5.3.11)$$

where l_2 is the length of the manipulator hand BC, and u_c is the displacement of the gripper C.

Loading the robot arm in a certain way, we can measure displacements at different points of the arm and then calculate the desired stiffness coefficients from Equations 5.3.5, 5.3.8, 5.3.10, and 5.3.11.

In our experiments, we used loads ranging from 0 to 63 N. The indicator head for measuring displacements was located in different positions shown in Figure 5.3.2. The measurements were carried out for the following three positions of the arm:

a. The arm is extended by its full length ($L - x = 500$ mm).
b. The first link is in the middle position ($L - x = 250$ mm).
c. The first link is completely hidden ($L - x = 0$).

Figure 5.3.2a shows arrangements for measuring total elastic displacements, due to compliances of the junctions and the link DA. When the junction O was rigidly supported, displacements were measured at the points shown in Figure 5.3.2b. To estimate the contribution of individual parts to the total displacement, we carried out experiments according to Figure 5.3.2c.

To investigate the elastic compliance and determine the bending line of the first link, DA, we supported this link at the junction O and applied the force $F = 63$ N to the point A. Vertical displacements of the link DA were measured over all the length of the link, at the points distanced by 50 mm.

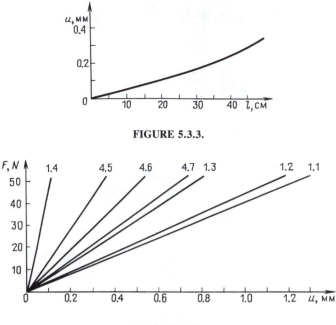

FIGURE 5.3.3.

FIGURE 5.3.4.

Having interpolated the measurement data, we obtained the shape of the bending curve (Figure 5.3.3).

Static experiments enabled us to determine elastic displacements at different points of the manipulator arm, depending on the loading weight and the chosen arrangement (see Figure 5.3.2). Some results for robot Ritm are presented graphically in Figure 5.3.4, where the loading force is drawn against elastic displacements. Each line is marked by two numbers. The first one indicates the arm extension and the position of a support, while the second corresponds to the location of the measuring head. These numbers match the respective numbers in Figure 5.3.2. The experiments also revealed residual displacements remaining after removing a load, and a hysteresis (see Figure 5.3.5).

The mean values of the stiffness coefficients c_O, c_A, and c_B, obtained from measurement data according to Equations 5.3.8, 5.3.10, 5.3.11, and 5.2.3, are equal to

$$\langle c_O \rangle = 8.85 \cdot 10^4 \ N{\cdot}m, \ \langle c_A \rangle = 2.94 \cdot 10^4 \ N{\cdot}m$$

$$\langle c_B \rangle = 0.72 \cdot 10^4 \ N{\cdot}m \tag{5.3.12}$$

Calculation of the standard deviations for these coefficients according to Equation 5.2.3 shows that c_O, c_A, and c_B are determined with the accuracy of 27, 7.2, and 15%, respectively.

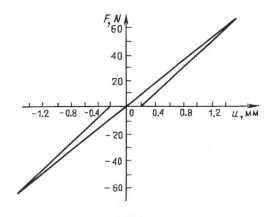

FIGURE 5.3.5.

Knowing the torsional stiffness coefficients c_O, c_A, and c_B, and also the bending stiffness of the link DA determined by Equation 5.3.5, we can calculate elastic displacements at any point of the manipulator under different loads and supports. Thus, we can verify our assumptions 1 through 3 made in this section by comparing computational and experimental data. For robot Ritm these data are in a good agreement.

The total displacement of the gripper is caused by the lumped elasticity of the junctions O, A, and B, and the distributed elasticity of the link DA. By experiments, we can estimate the relative influence of these components. For this purpose, we measure displacements due to the flexibility of separate parts (supporting the arm at proper places) and the total displacement of the gripper. Since the total displacement is the sum of partial ones, in the linear approximation, we can readily estimate the relative contributions. For robot Ritm with the fully extended arm, the contributions are as follows:

1. The elasticity of the junction O gives 57% of the total compliance.
2. The elasticity of the link DA gives 26%.
3. The elasticity of the junction B gives 13%.
4. The elasticity of the junction A gives 4%.

It is worth noting that the overall contribution of the junctions is considerably greater than the contribution of the link. This is typical for industrial robots, and when compiling a mathematical model of a robot we often can neglect the flexibility of links without essential loss of accuracy. Thus, to increase the accuracy of industrial robots by reducing their elastic compliance, we must make joints and junctions stiffer. This can be achieved by using new materials, in particular composites, as well as by perfecting the structural design of joints and junctions.

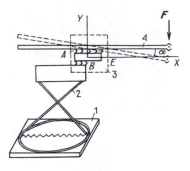

FIGURE 5.4.1.

5.4. ELASTIC COMPLIANCE OF AN ELECTROMECHANICAL ROBOT

In Figure 5.4.1, the kinematic structure of electromechanical industrial robot Universal-5 is presented. Robots of such type are widely used in different applications. These robots consist of a fixed platform (1); a unit (2) that provides lifting and lowering of the robot body, as well as its rotation about the vertical axis; a junction (3) that admits extension of the arm and its rotation (about the vertical axis) relative to the body; and an arm (4) with a gripper. Arms of these robots are usually made of comparatively thick tubes. Such arms are rather stiff and can be regarded as rigid bodies.

Let us connect the coordinate frame OXY with the junction (3). Here, the Y axis coincides with the rotation axis of the arm, and the X axis is directed along the arm. Denote by c_O the total stiffness of the junction (3), by c_1 and c_2 the stiffnesses of the junctions A and B incorporated in the junction (3), by l the length of the manipulator arm from the point O to the gripper, by L the distance from the Y axis to the mass center of the arm, and by α the angular elastic displacement of the arm under the action of the vertical force F applied to the gripper. The elastic displacements are considered only in the vertical plane.

When the force F acts on the gripper and the manipulator is in equilibrium, the torque equal to $c_O \alpha = Fl$ acts at the joint (3). Therefore, we have

$$c_O = Fl^2/u \qquad (5.4.1)$$

where u is the elastic displacement of the gripper under the action of the force F. Loading the gripper by a force F, we measure directly the displacement u and calculate the total stiffness c_O of the junction 3 using Equation 5.4.1. We can also determine partial stiffnesses c_1 and c_2. To obtain c_1, we support the arm at the point E (see Figure 5.4.1) so as to eliminate the influence of

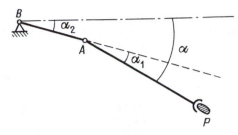

FIGURE 5.4.2.

the junction B, and repeat the procedure for finding the stiffness c_o. The stiffness c_2 is evaluated as follows:

$$c_2 = \frac{c_o c_1}{c_1 - c_o} \tag{5.4.2}$$

To derive Equation 5.4.2, consider a mechanical model of the junction (3). The arm fastened at the junction (3) can be regarded as a two-member struc-ture, *BAP*, with elastic joints (see Figure 5.4.2). The joint B, with the stiffness c_2, corresponds to the junction B, while the joint A, with the stiffness c_1, models the junction A (see Figure 5.4.1). We denote by α_1 and α_2 angular displacements at the joint A and B, respectively, while $\alpha = \alpha_1 + \alpha_2$ is the total angle.

Let us express the total stiffness c_o of the junction (3) in terms of the partial stiffnesses c_1 and c_2. The elastic potential energy of the system is given by

$$\Pi = \frac{1}{2}(c_1\alpha_1^2 + c_2\alpha_2^2) \tag{5.4.3}$$

When the system is in equilibrium with a given α, potential energy (Equation 5.4.3) attains its minimum under the constraint $\alpha_1 + \alpha_2 = \alpha$. Eliminating $\alpha_2 = \alpha - \alpha_1$ from Equation 5.4.3, we obtain

$$\Pi = \frac{1}{2}[c_1\alpha_1^2 + c_2(\alpha - \alpha_1)^2] \tag{5.4.4}$$

Minimizing Π in Equation 5.4.4 with respect to α_1, we find

$$\alpha_1 = \frac{c_2\alpha}{c_1 + c_2}$$

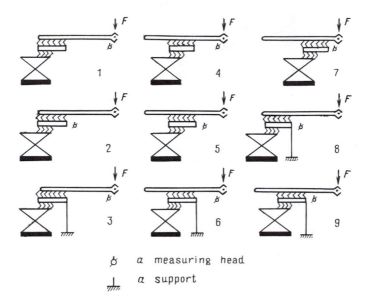

ϕ a measuring head

$\perp$ a support

FIGURE 5.4.3.

Inserting α_1 into Equation 5.4.4, we get

$$\Pi = \frac{1}{2} \frac{c_1 c_2}{c_1 + c_2} \alpha^2$$

and, therefore,

$$c_O = \frac{c_1 c_2}{c_1 + c_2} \tag{5.4.5}$$

Solving Equation 5.4.5 for c_2, we arrive at Equation 5.4.2.

In the course of experiments, we measured the vertical displacement of the gripper, depending on the load. The weight varied from 0 (unloaded gripper) to 98 N. In Figure 5.4.3, different arrangements for the measurements are shown. We made the measurements for three positions of the manipulator arm: the arm is fully moved out ($l = 840$ mm, positions 1, 2, 3, and 8 in Figure 5.4.3), the arm is half-extended ($l = 420$ mm, positions 4, 5, 6, and 9), and the arm is fully moved in ($l = 0$, position 7). Measurements with a support (positions 3, 6, 8, and 9 in Figure 5.4.3) enabled us to determine the stiffness c_1 of the junction A (see Figure 5.4.1). We also verified experimentally the assumption that the arm is practically rigid, whereas the elastic compliance is lumped at the junctions. We placed a certain load ($F = 98$ N) in the gripper and measured the vertical displacement of the fully extended

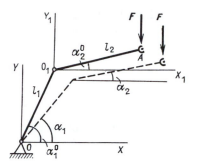

FIGURE 5.5.1.

arm at different points distanced by 20 mm from each other. The bending curve turned out to be a straight line, within the measurement accuracy. As a result of our experiments, we obtained mean values for the stiffnesses c_O and c_1:

$$\langle c_O \rangle = 20 \cdot 10^3 \text{ N·m}, \ \langle c_1 \rangle = 33 \cdot 10^3 \text{ N·m}$$

Calculation of the standard deviations showed that c_O and c_1 were determined with accuracy up to 15 and 4%, respectively. Knowing c_O and c_1, we can estimate the stiffness c_2 of the junction B by using Equation 5.4.2.

5.5. ELASTIC COMPLIANCE OF AN ANTHROPOMORPHIC INDUSTRIAL ROBOT

Consider an anthropomorphic manipulation robot with the kinematic structure shown in Figure 5.5.1. The manipulator consists of two links, OO_1 and O_1A, whose lengths are equal to l_1 and l_2, respectively. The links are regarded as rigid bars connected by revolute elastically compliant joints O and O_1. The plane OO_1A of the manipulator is vertical. Deformation of the manipulator is caused by its own weight and the weight of a load in the gripper.

We introduce two coordinate frames, OXY and $O_1X_1Y_1$, whose origins O and O_1 are placed at the respective joints, the axes OX and O_1X_1 are horizontal, while the axes OY and O_1Y_1 are oriented vertically. Denote by α_1 the angle between the link OO_1 and the axis OX, and by α_2 the angle between the link O_1A and the axis O_1X_1. For the unloaded manipulator (no load in the gripper), we denote these angles by α_1^0 and α_2^0, respectively.

Denote by $\mathbf{r}$ the radius-vector of the gripper (the point A) with respect to the point O, by $\mathbf{r}_0$ the radius-vector $\mathbf{r}$ for the unloaded manipulator, by $\Delta\mathbf{r} = \mathbf{r} - \mathbf{r}_0$ the elastic displacement of the gripper caused by the load weight F, by c_1 and c_2 the stiffnesses of the joints O and O_1, respectively, and by φ_1 and φ_2 the additional rotation angles of the manipulator links OO_1 and

O_1A (respectively) due to the load in the gripper. The angles α_i and $\varphi_i (i = 1,2)$ are related by

$$\alpha_1 = \alpha_1^0 + \varphi_1, \quad \alpha_2 = \alpha_2^0 + \varphi_2 \tag{5.5.1}$$

We assume that the stiffnesses c_1 and c_2 of the manipulator joints are large so Equations 4.2.1 and 4.2.2 hold. Therefore, to describe the equilibrium state, we can use linear Equations 5.2.2 and kinematic Equations 5.2.1. For our case, we have $N = 1$, $m = 2$, $\alpha = (\alpha_1^0, \alpha_2^0)$, $u = (\varphi_1, \varphi_2)$, and $r_1 = r$. Then Equation 5.2.1 becomes

$$\mathbf{r} = \begin{Vmatrix} l_1\cos \alpha_1 + l_2\cos \alpha_2 \\ l_1\sin \alpha_1 + l_2\sin \alpha_2 \end{Vmatrix}$$

$$= \begin{Vmatrix} l_1\cos(\alpha_1^0 + \varphi_1) + l_2\cos(\alpha_2^0 + \varphi_2) \\ l_1\sin(\alpha_1^0 + \varphi_1) + l_2\sin(\alpha_2^0 + \varphi_2) \end{Vmatrix} \tag{5.5.2}$$

Equation 5.5.2 represents the radius-vector $\mathbf{r}$ of the gripper in the reference frame *OXY*. The potential energy of elastic deformations caused by the load in the gripper is given by

$$\Pi = \frac{1}{2}[c_1\varphi_1^2 + c_2(\varphi_2 - \varphi_1)^2] = \frac{1}{2}[(c_1 + c_2)\varphi_1^2 - 2c_2\varphi_2\varphi_1 + c_2\varphi_2^2]$$

$$= \frac{1}{2}\sum_{i,j=1}^{2} c_{ij}\varphi_i\varphi_j$$

Here, c_{ij} are the elements of the stiffness matrix C (see Equation 5.2.2) equal to

$$c_{11} = c_1 + c_2, \; c_{12} = c_{21} = -c_2, \; c_{22} = c_2$$

The force acting on the gripper is the load weight. Hence, in Equation 5.2.2 we have

$$\mathbf{F}_1 = \|0, \, -F\| \tag{5.5.3}$$

Substituting $\Pi_1 = \Pi$ and $\mathbf{F}_1$ from Equations 5.5.2, 5.5.3, and $u = (\varphi_1, \varphi_2)$ into Equation 5.2.2, we obtain the equilibrium equations:

$$-Fl_1\cos \alpha_1^0 = c_1\varphi_1 - c_2(\varphi_2 - \varphi_1)$$

$$-Fl_2\cos \alpha_2^0 = c_2(\varphi_2 - \varphi_1)$$

Solving these equations for c_1 and c_2 gives

$$c_1 = -\frac{F(l_1\cos\alpha_1^0 + l_2\cos\alpha_2^0)}{\varphi_1}, \quad c_2 = -\frac{Fl_2\cos\alpha_2^0}{\varphi_2 - \varphi_1} \qquad (5.5.4)$$

Linearizing Equation 5.5.2 in φ_1 and φ_2, we get

$$\Delta\mathbf{r} = \mathbf{r} - \mathbf{r}_0 = \left\|\begin{matrix}\Delta x \\ \Delta y\end{matrix}\right\| = \left\|\begin{matrix}-l_1\varphi_1\sin\alpha_1^0 - l_2\varphi_2\sin\alpha_2^0 \\ l_1\varphi_1\cos\alpha_1^0 + l_2\varphi_2\cos\alpha_2^0\end{matrix}\right\|$$

$$(5.5.5)$$

Here, Δx and Δy are horizontal and vertical displacements of the gripper, respectively. If $\sin(\alpha_2^0 - \alpha_1^0) \neq 0$, Equations 5.5.5 can be solved with respect to φ_1, φ_2:

$$\varphi_1 = \frac{\Delta x \cos\alpha_2^0 + \Delta y \sin\alpha_2^0}{l_1\sin(\alpha_2^0 - \alpha_1^0)}$$

$$\varphi_2 = -\frac{\Delta x \cos\alpha_1^0 + \Delta y \sin\alpha_1^0}{l_2\sin(\alpha_2^0 - \alpha_1^0)} \qquad (5.5.6)$$

Substituting l_1 and l_2 from Equations 5.5.6 into 5.5.4, we obtain the joint stiffnesses c_1 and c_2 expressed through horizontal and vertical displacements of the gripper:

$$c_1 = \frac{Fl_1 \sin(\alpha_1^0 - \alpha_2^0)\cdot(l_1\cos\alpha_1^0 + l_2\cos\alpha_2^0)}{\Delta x \cos\alpha_2^0 + \Delta y \sin\alpha_2^0}$$

$$c_2 = \frac{Fl_1 l_2^2\cos\alpha_2^0 \sin(\alpha_2^0 - \alpha_1^0)}{\Delta x(l_1\cos\alpha_1^0 + l_2\cos\alpha_2^0) + \Delta y(l_1\sin\alpha_1^0 + l_2\sin\alpha_2^0)} \qquad (5.5.7)$$

The horizontal, Δx, and vertical, Δy, displacements caused by the force F are directly measured in static experiments. Then the stiffnesses of the joints are calculated according to Equations 5.5.7. We carried out such experiments with anthropomorphic electromechanical robot RPM-25 (see Section 1.1.3). The lengths of the links of this robot are equal to $l_1 = 700$ mm and $l_2 = 1500$ mm.

The measurements were made for different configurations of the arm and with different locations of indicator heads. These arrangements are shown in Figure 5.5.2. The weight of the load in the gripper was varied from 0 to 294 N. In all configurations, the angle α_2^0 was equal to zero, while α_1^0 was set at 90° (cases 1 to 3 in Figure 5.5.2), 35° (cases 4 to 6), and 145° (cases 7 to 9). In cases 2, 5, and 8, we used a support to eliminate the elastic compliance of the first joint and to evaluate the influence of the flexibility of the second

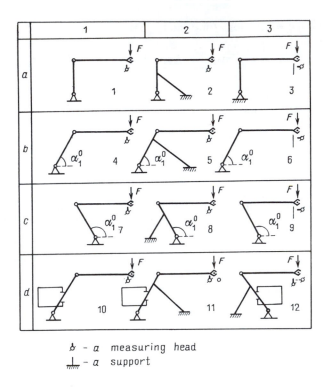

$♭$ - a measuring head
$⊥$ - a support

FIGURE 5.5.2.

joint only. In robot RPM-25, the rotation angles of the links are restricted by mechanical limit stops that have their own elastic compliance. The flexibility of the stops was also studied experimentally (cases 10 to 12 in Figure 5.5.2). We set the link in a limit position, measured elastic displacements of the gripper under a specified load, and determined the total stiffness of the joint and the stop by using Equations 5.5.7. Comparing the total stiffness with the stiffness of the joint, we estimated the stiffness of the limit stop.

We also verified experimentally the assumption that the links are practically rigid and showed that the elastic compliance of the robot is concentrated mostly at its joints. For example, for the second link it was done as follows. We fixed the first link using a support, as shown in Figure 5.5.2 (case 2), loaded the gripper by the weight $F = 294$ N, and measured vertical displacements at different points of the second link distanced by 50 mm from each other. It turned out that the displacements linearly depended on the distance from the second joint, and therefore the shape of the link was rectilinear.

As a result of the experiments, we have obtained the elastic displacements of the gripper against the weight of the load. The corresponding graphs are shown in Figure 5.5.3 by solid lines, for cases 1, 2, and 3 (see Figure 5.5.2). For other cases the qualitative picture is similar.

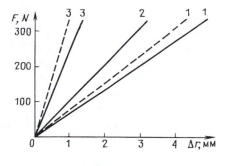

FIGURE 5.5.3.

The experimental results can be used for determining the joint stiffnesses according to Equations 5.5.7. For configuration a in Figure 5.5.2, Equations 5.5.7 look as follows:

$$c_1 = Fl_1l_2/\Delta x$$

$$c_2 = Fl_2^2l_1(|\Delta y|l_1 - l_2\Delta x)^{-1}, \Delta y < 0$$

For configuration b, we have

$$c_1 = Fl_1\sin 35°(l_1\cos 35° + l_2)/\Delta x$$

$$c_2 = Fl_2^2l_1\sin 35°[|\Delta y|l_1\sin 35° - \Delta x(l_1\cos 35° + l_2)]^{-1}$$

For configuration c, Equations 5.5.7 take the form

$$c_1 = Fl_1\sin 35°(l_2 - l_1\cos 35°)/\Delta x$$

$$c_2 = Fl_2^2l_1\sin 35°[|\Delta y|l_1\sin 35° - \Delta x(l_2 - l_1\cos 35°)]^{-1}$$

The mean values (over all experiments) for the stiffness coefficients c_1 and c_2 appeared to be the following:

$$c_1 \approx 32 \cdot 10^4 \text{ N·m}, c_2 \approx 39 \cdot 10^4 \text{ N·m} \tag{5.5.8}$$

The estimation accuracy makes up 18% for c_1 and 16% for c_2.

Having known c_1 and c_2, we can calculate the elastic displacements for different configurations of the manipulator and compare the results with the experimental data.

Solving Equations 5.5.7 for Δx and Δy we obtained

$$\Delta x = F[l_1\cos \alpha_1^0 + l_2\cos \alpha_2^0)$$

$$\times (l_1\sin \alpha_1^0 + l_2\sin \alpha_2^0)/c_1 + l_2^2\sin \alpha_2^0\cos \alpha_2^0/c_2]$$

$$\Delta y = -F[l_2^2\cos^2 \alpha_2^0/c_2 + (l_1\cos \alpha_1^0 + l_2\cos \alpha_2^0)^2/c_1] \tag{5.5.9}$$

The results of calculations according to Equations 5.5.9 are shown by dashed lines in Figure 5.5.3, for configuration a (see Figure 5.5.2). For other configurations the qualitative picture is similar. These results enable us to conclude about the adequacy of the adopted mechanical model and the accuracy of determining the stiffness coefficients c_1 and c_2.

Analysis of the experimental results shows that individual joints make different contributions to the total elastic displacement of the gripper of robot RPM-25. Moreover, the relative contributions of the joints depend on the manipulator configuration. It turned out that:

1. For configuration a, the contribution of the first joint makes up 26%, while the contribution of the second joint is 74%.
2. For configuration b, the contributions of the first and second joints to the total compliance are 30 and 70%, respectively.
3. For configuration c, the contributions of the first and second joints make up 19 and 81%, respectively.

Using Equations 5.5.9, we can obtain similar estimates for any configurations of the manipulator (for arbitrary α_1^0 and α_2^0).

5.6. RESULTS OF DYNAMIC TESTING OF INDUSTRIAL ROBOTS

In Section 5.4 we described the results of static experiments with industrial robot Universal-5. The static experiments showed that the elastic compliance of robot Universal-5 was lumped at junction (3) (see Figure 5.4.1). The elasticity of junction (3) gives rise to additional degrees of freedom for the gripper motion in the vertical and horizontal planes.

Let us consider the elastic oscillations of the gripper arising when the arm moves vertically or rotates about the vertical axis. We denote the additional (caused by the elastic compliance) angles of the arm rotation in the vertical and horizontal planes by α and β, respectively.

The kinetic energy and the elastic potential energy can be written as follows:

$$T = 1/2(A\dot{\psi}, \dot{\psi}), \quad \Pi = 1/2(C\psi, \psi)$$

$$\psi = \left\| \begin{matrix} \alpha(t) \\ \beta(t) \end{matrix} \right\|, \quad A = \left\| \begin{matrix} I & 0 \\ 0 & I \end{matrix} \right\|, \quad C = \left\| \begin{matrix} c & 0 \\ 0 & c \end{matrix} \right\| \qquad (5.6.1)$$

We assume that the stiffness coefficients of junction (3) have the same values for rotations in the horizontal and vertical planes. Here, $I = m(l^2/12 + L^2)$ is the moment of inertia of the arm about the rotation axis (OY or OZ; see Figure 5.4.1), l is the length of the arm, L is distance between the center of mass of the arm and the axis OY.

The arm of the manipulator is a steel tube of annular cross section. The length of the arm is $l = 840$ mm; the distance between the center of mass

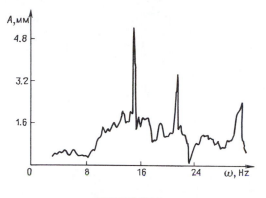

FIGURE 5.6.1.

and the rotation axis is $L = 420$ mm (when the arm is completely stretched); the outer and inner diameters of the tube are equal to $D_1 = 100$ mm and $D_2 = 20$ mm, respectively. The mass of the arm is $m = 12$ kg, its moment of inertia about the rotation axis OY is $I = 28$ N·m².

The eigenfrequencies of elastic oscillations of the arm are determined from the equation $\det(C - \omega^2 A) = 0$. In our particular case, this equation takes the form $\omega_{1,2}^2 = c/I$. The theoretical calculations give $\omega_{1,2} = 13$ Hz for the completely stretched arm, and $\omega_{1,2} = 27$ Hz for the half-stretched arm. Here, c has been determined from the static experiments (see Section 5.4).

The results of dynamic experiments show that the low resonant frequencies of the elastic oscillations of the gripper of robot Universal-5 coincide with the above mentioned theoretically predicted frequencies, within the accuracy of 3 to 5%. A picture of the spectral characteristic of the arm is given in Figure 5.6.1. In this experiment, the arm was completely stretched and the gripper was unloaded. According to Figure 5.6.1, the arm has resonant frequencies (eigenfrequencies) equal to $\omega_1 = 14.5$ Hz, $\omega_2 = 20$ Hz, and $\omega_3 = 31$ Hz. For the lowest (basic) mode of the elastic oscillations, we have a good agreement between the results of the dynamic experiments and the calculations according to the formula $\omega = (c/I)^{1/2} \approx 13$ Hz, where the stiffness coefficient c results from the static experiments. In case the arm is extended by 20% of its maximal length, the eigenfrequencies become 1.4 times as large as for the completely stretched arm. Apart from the eigenfrequencies of the elastic oscillations of the gripper, the spectrum also contains the frequencies of rapidly decaying oscillations, due to the elasticity of the lifting mechanism of the robot, the compliance of its platform, and oscillatory properties of the control system.

Dynamic experiments showed that the elastic oscillations with resonant frequencies were more intensive at the beginning and the end of the motion of the robot, when it accelerates or decelerates. It is typical for electromechanical manipulators as well as for robots with pneumatic and hydraulic drives.

Chapter 6

DYNAMICS OF INDUSTRIAL ROBOTS WITH ELASTIC JOINTS

6.1. PROBLEMS OF DYNAMICS FOR ROBOTS WITH DIFFERENT KINEMATIC STRUCTURES

6.1.1. MECHANICAL MODELS

Consider two models of industrial manipulation robots. The first one is a two-member anthropomorphic manipulator (Figure 6.1.1) consisting of the link (1) (O_1O_2) connected with a fixed base by a revolute joint O_1, and the link (2) (O_2P) with a gripper. The manipulator links are rigid bars connected to each other by a revolute joint, O_2. The robot is controlled by two independent drives located at the joints O_1 and O_2 and generating the torques M_1 and M_2 with respect to the joint axes. The motion of the system occurs in the vertical plane. To describe the motion, we introduce two coordinate frames $O_1X_1Y_1Z_1$ and $O_2X_2Y_2Z_2$ (see Figure 6.1.1). The frame $O_1X_1Y_1Z_1$ is connected with the fixed base. The axis O_1Z_1 is horizontal and directed along the axis of the joint O_1. The axes O_1X_1 and O_1Y_1 lie in the plane of the manipulator arm O_1O_2P. The axis O_1X_1 is horizontal, while O_1Y_1 is directed vertically. The axes of the frame $O_2X_2Y_2Z_2$ are parallel to the axes of the frame $O_1X_1Y_1Z_1$.

The generalized coordinates are the angles q_1 and q_2 formed by the links 1 and 2, respectively, with the horizontal axis O_1X_1; the angles φ and ψ formed by the links 1 and 2 with the horizontal axis in the absence of the elastic compliance of the joints; and the complementary angles $\alpha_1 = q_1 - \varphi$ and $\beta_1 = q_2 - \psi$ describing the elastic displacements.

Denote by m_1 and m_2 the masses of the links 1 and 2; by I_1 and I_2, their inertia moments about the axes of the joints O_1 and O_2, respectively; by L_1 and L_2, the distances from the axes of the joints O_1 and O_2 to the mass centers of the corresponding links; by l_1, the length of the first link ($l_1 = |O_1O_2|$); by c_1 and c_2, the stiffness coefficients of the joints O_1 and O_2; and by g, the gravity acceleration.

The second model (Figure 6.1.2) is a cylindrical coordinate manipulation robot consisting of the fixed base (1), the vertical column (2), and the arm (3) with the gripper. The arm is a rectilinear homogeneous bar of a constant annular cross section. The base, the column, and the arm are regarded as rigid bodies. The arm is connected to the column by a revolute joint possessing the elastic compliance. The manipulator has three controlled degrees of freedom corresponding to the vertical motion of the column (lifting and lowering the arm), translation of the arm along the horizontal guide, and its rotation about the vertical axis. The robot is controlled by means of the torque M with respect to the arm rotation axis, the vertical force F_1 applied to the

149

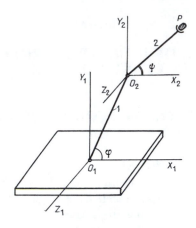

FIGURE 6.1.1.

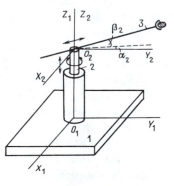

FIGURE 6.1.2.

column, and the force F_2 applied to the arm and directed along the guide. To describe the motion of the manipulator shown in Figure 6.1.2, we introduce two coordinate frames: the fixed coordinate frame $O_1X_1Y_1Z_1$ connected with the base, and the rotating frame $O_2X_2Y_2Z_2$ attached to the manipulator arm. The axes O_1Z_1 and O_2Z_2 coincide with the axis of the arm rotation. The axis O_2Y_2 is oriented along the arm, in case the displacements due to elasticity of the joint O_2 are absent.

We take the following variables as the generalized coordinates: the distance x from the joint O_2 to the gripper; the distance z between the coordinate planes $O_1X_1Y_1$ and $O_2X_2Y_2$ (the height of the arm lifting); the angle q formed by the projection of the arm onto the plane $O_2X_2Y_2$ with the horizontal axis O_1Y_1 (the total rotation angle of the arm); the angle χ between the axes O_1Y_1 and O_2Y_2 (the rotation angle of the arm in the absence of elastic displacements); the angle $\alpha_2 = q - \chi$ formed by the projection of the manipulator arm onto the plane $O_2X_2Y_2$ with the axis O_2Y_2; and the angle β_2 between the manipulator arm and the plane $O_2X_2Y_2$. The angels α_2 and β_2 correspond to the additional degrees of freedom due to the elastic compliance of the joint O_2.

Denote by l the length of the manipulator arm; by L, the coordinate of the projection of the mass center of the arm onto the axis O_2Y_2 in the absence of elastic displacements ($|L|$ is the distance between the joint O_2 and the mass center of the arm); by m, the mass of the arm; by m', the mass of the column; by I_0, the inertia moment of the arm about the axis passing through the mass center normally to the axis of the arm; and by I, the inertia moment of the arm about the axis passing through the point O_2 normally to the axis of the arm. According to the Huygens-Steiner theorem (see Section 2.3), $I = I_0 + mL^2$. The variables L, l, and x are related by $L = x - l/2$.

The kinetic energies of two mechanical systems described above are given by

$$T_1 = \frac{1}{2}(I_1 + m_2 l_1^2)\dot{q}_1^2 + \frac{1}{2}I_2\dot{q}_2^2 + m_2 l_1 L_2 \cos(q_1 - q_2)\dot{q}_1\dot{q}_2$$

$$q_1 = \varphi + \alpha_1, \; q_2 = \psi + \beta_1$$

$$T_2 = \frac{1}{2}\{(m + m')\dot{z}^2 + m[L^2(\dot{\chi} + \dot{\alpha}_2)^2 \cos^2\beta_2$$

$$+ \dot{L}^2 + L^2\dot{\beta}^2 + 2\dot{\beta}_2\dot{z}L\cos\beta_2 + 2L\dot{z}\sin\beta_2]$$

$$+ I_0\dot{\beta}^2 + I_0(\dot{\chi} + \dot{\alpha}_2)^2\cos^2\beta^2\} \tag{6.1.1}$$

Here and henceforth, T_1 is related to the manipulator shown in Figure 6.1.1, whereas T_2 corresponds to the model given in Figure 6.1.2.

The respective potential energies Π_i, $i = 1, 2$ are sums of the potential energies of gravity forces (Π'_i) and the elastic energies due to the joint deformations:

$$\Pi_i = \frac{1}{2}(C_i\Phi_i, \Phi_i) + \Pi'_i$$

$$C_1 = \left\|\begin{matrix} c_1 + c_2 & -c_2 \\ -c_2 & c_2 \end{matrix}\right\|, \quad C_2 = \left\|\begin{matrix} c & 0 \\ 0 & c \end{matrix}\right\|, \quad \Phi_i = \left\|\begin{matrix} \alpha_i \\ \beta_i \end{matrix}\right\|,$$

$$i = 1, 2$$

$$\Pi'_1 = m_1 g L_1 \sin(\alpha_1 + \varphi) + m_2 g[l_1\sin(\alpha_1 + \varphi) + L_2\sin(\beta_1 + \psi)]$$

$$\Pi'_2 = (m + m')gz + mgL\sin\beta_2 \tag{6.1.2}$$

The virtual work δA_i ($i = 1, 2$) carried out by control forces is given by

$$\delta A_1 = (M_1 - M_2)\delta q_1 + M_2\delta q_2 - (M_1 - M_2)\delta\alpha_1 - M_2\delta\beta_1$$

$$= (M_1 - M_2)\delta\varphi + M_2\delta\psi$$

$$\delta A_2 = (F_1 + F_2\sin\beta_2)\delta z + F_2\delta L + M\delta\chi \tag{6.1.3}$$

The experimental investigations described in Chapter 5 show that for industrial manipulation robots, elastic displacements are small compared with linear dimensions of the links, for all admissible loads. Therefore, the vectors $\Phi_i = \|\alpha_i, \beta_i\|^T$ and the matrices C_i, $i = 1, 2$ can be represented as follows:

$$\Phi_i = \epsilon^2\Phi_{i*}, \qquad C_i = \epsilon^{-2}C_{i*}, \qquad \Phi_{i*} \sim 1$$

$$C_{i*} \sim 1, \qquad \epsilon \ll 1 \tag{6.1.4}$$

6.1.2. KINEMATIC CONTROL

Consider the following problem.

Problem 6.1. Let the time histories of the generalized coordinates,

$$\varphi = \varphi(t), \qquad \psi = \psi(t)$$

$$\text{or} \qquad \chi = \chi(t), \qquad z = z(t), \qquad L = L(t) \tag{6.1.5}$$

be specified for the robots shown in Figure 6.1.1 or 6.1.2, respectively. Find the motion of the robots, taking into account the elastic compliance of their joints, i.e., find the functions $\alpha_i(t)$ and $\beta_i(t)$, $i = 1, 2$, as well as the controls $(M_1(t), M_2(t))$ or $(M(t), F_1(t), F_2(t))$ providing the time histories given by Equation 6.1.5.

Problem 6.1 is a problem of kinematic control (see Problem 3.1) for the robots shown in Figures 6.1.1 and 6.1.2.

The Lagrangian equations corresponding to the generalized coordinates α_i and β_i for the mechanical systems described by Equations 6.1.1 to 6.1.4 can be represented by

$$\epsilon^2 \frac{d}{dt} [A_i(t)\dot{\Phi}_{i*}] + [C_{i*} + \epsilon^2 D_i(t)]\Phi_{i*}$$

$$- \epsilon^2 G_i(t)\dot{\Phi}_{i*} + B_i(t) + O(\epsilon^2) = 0, \qquad i = 1, 2$$

$$A_1 = \begin{Vmatrix} I_1 + m_2 l_1^2 & m_2 l_1 L_2 \cos \delta \\ m_2 l_1 L_2 \cos \delta & I_2 \end{Vmatrix}, \qquad \delta = \varphi - \psi$$

$$B_1 = \begin{Vmatrix} (I_1 + m_2 l_1^2)\ddot{\varphi} + m_2 l_1 L_2(\ddot{\psi}\cos\delta + \dot{\psi}^2\sin\delta) \\ \quad + g(m_1 L_1 + m_2 l_1)\cos\varphi \\ I_2\ddot{\psi} + m_2 l_1 L_2(\ddot{\varphi}\cos\delta - \dot{\varphi}^2\sin\delta) + m_2 g L_2\cos\psi \end{Vmatrix}$$

$$D_1 = \begin{Vmatrix} -m_2 l_1 L_2(\ddot{\psi}\sin\delta - \dot{\psi}^2\cos\delta) & m_2 l_1 L_2(\ddot{\psi}\sin\delta - \dot{\psi}^2\cos\delta) \\ \quad - g(m_1 L_1 + m_2 l_1)\sin\varphi & \\ -m_2 l_1 L_2(\ddot{\varphi}\sin\delta + \dot{\varphi}^2\cos\delta) & m_2 l_1 L_2(\ddot{\varphi}\sin\delta + \dot{\varphi}^2\cos\delta) \\ & \quad - m_2 g L_2\sin\psi \end{Vmatrix}$$

$$G_1 = m_2 l_1 L_2 \sin\delta \begin{Vmatrix} 0 & -(\dot{\varphi} + \dot{\psi}) \\ \dot{\varphi} + \dot{\psi} & 0 \end{Vmatrix}$$

$$A_2 = \begin{Vmatrix} I & 0 \\ 0 & I \end{Vmatrix}, \qquad B_2 = \begin{Vmatrix} I\ddot{\chi} + 2mL\dot{L}\dot{\chi} \\ mL\ddot{z} + mgL \end{Vmatrix}, \qquad I = I_0 + mL^2$$

$$D_2 = \begin{Vmatrix} 0 & 0 \\ 0 & I\dot{\chi}^2 \end{Vmatrix}, \qquad G_2 = 0 \tag{6.1.6}$$

Following the technique developed in Chapter 3 (Section 3.2), we seek the solution of Equations 6.1.6 as the sum

$$\Phi_{i*}(t) = \Phi_i^0(t) + \theta_i(\tau), \qquad \tau = \epsilon^{-1}t, \qquad i = 1, 2 \tag{6.1.7}$$

where τ is the fast time. The term $\Phi_i^0(t)$ describes the slowly changing (quasistatic) displacements. The characteristic time of these displacements is of the order O(1), i.e., of the order of the time required for the robot to implement its operation. The second term in Equation 6.1.7 describes fast elastic oscillations of the manipulator with frequencies of the order of ϵ^{-1}.

Substituting $\phi_i(t)$ from Equation 6.1.7 into Equations 6.1.6 and omitting the terms of the order $O(\epsilon^2)$, we obtain

$$\left\{ \frac{d}{d\tau}\left[A_i(t)\frac{d\theta_i(t)}{d\tau} \right] + C_{i*}\theta_i(\tau) - \epsilon G_i(t)\frac{d\theta_i(\tau)}{d\tau} \right\}$$
$$+ \{ C_{i*}\Phi_i^0(t) + B_i(t) \} = 0, \qquad i = 1, 2 \tag{6.1.8}$$

There is an arbitrariness in representing the solution in the form of Equation 6.1.7. The arbitrariness disappears if we require that each expression in curly brackets in Equation 6.1.8 should be zero. This requirement stems from the fact that for the manipulator with rigid joints ($\epsilon \rightarrow 0$), the equation of motion is

$$C_{i*}\Phi_i^0(t) + B_i(t) = 0, \qquad i = 1, 2 \qquad (6.1.9)$$

The left-hand side of this equation is just the expression in the second curly brackets. For more detail, see Chapter 3, Section 3.1.

From Equation 6.1.9, we obtain the quasistatic displacement:

$$\Phi_i^0(t) = -C_{i*}^{-1}B_i(t), \qquad i = 1, 2 \qquad (6.1.10)$$

In the case of kinematic control, where functions in Equations 6.1.5 are specified, the vector $B_i(t)$ is a known function of time (see Equations 6.1.6). The stiffness matrix C_i can be determined experimentally, for example, by the technique of Chapter 5.

Equating to zero the expression in the first curly brackets in Equation 6.1.8, we get

$$\frac{d}{d\tau}\left[A_i(t)\frac{d\theta_i}{d\tau}\right] + C_{i*}\theta_i = \epsilon G_i(t)\frac{d\theta_i(t)}{d\tau}, \qquad i = 1, 2 \quad (6.1.11)$$

This system of differential equations with slowly changing parameters is analogous to Equation 3.2.2 and governs fast elastic oscillations of the manipulator. The approximate solution of Equation 3.2.2 obtained by the averaging method is given by Equations 3.2.5 to 3.2.10. We obtain from these formulas (taking into account that the matrices G_i from Equations 6.1.6 are skew-symmetric) the general solution of Equation (6.1.11) in the first approximation:

$$\theta_i(\tau) = \sum_{k=1}^{2} X_{ik}(t)d_{ik}(t)\cos\gamma_{ik}$$

$$d_{ik}(t) = d_{ik}^0\left[\frac{m_{ik}(0)\omega_{ik}(0)}{m_{ik}(t)\omega_{ik}(t)}\right]^{1/2}$$

$$\gamma_{ik} = \epsilon^{-1}\int_0^t \omega_{ik}(\xi)d\xi + \gamma_{ik}^0$$

$$m_{ik}(t) = (X_{ik}, A(t)X_{ik}), \qquad i, k = 1, 2 \qquad (6.1.12)$$

Here, ω_{ik} are positive roots of the characteristic equations

$$\det[C_{i*} - \omega_{ik}^2 A_i(t)] = 0, \qquad i, k = 1, 2 \qquad (6.1.13)$$

The quantities $\omega_{ik}(t)$ can be interpreted as instantaneous (for current t) eigen-frequencies of elastic oscillations. By $X_{ik}(t)$ we denote eigenvectors of these oscillations, i.e., nonzero solutions of the system of algebraic equations:

$$[C_{i*} - \omega_{ik}^2(t)A_i(t)]X_{ik} = 0, \qquad i, k = 1, 2$$

The variables $d_{ik}(t)$ and γ_{ik} in Equations 6.1.12 are the amplitude and phase of the elastic oscillations of the manipulator. Let us recall that the index i shows the manipulator model, while the index k corresponds to the number of the natural oscillation mode.

Constants d_{ik}^0 and γ_{ik}^0 in Equations 6.1.12 can be determined from initial conditions. Let the angles of elastic displacement, $\Phi_i(0)$, and the rates of their change, $\dot{\Phi}_i(0)$, be specified for $t = 0$. Then, using Equations 6.1.4, 6.1.7, and 6.1.12, we obtain the system of algebraic equations for the parameters d_{ik}^0 and γ_{ik}^0:

$$\Phi_i^0(0) + \sum_{k=1}^{2} X_{ik}(0)d_{ik}^0 \cos \gamma_{ik}^0 = \epsilon^{-2}\Phi_i(0)$$

$$-\sum_{k=1}^{2} X_{ik}(0)\omega_{ik}(0)d_{ik}^0 \sin \gamma_{ik}^0 = \epsilon^{-1}\dot{\Phi}_i(0) \qquad (6.1.14)$$

In the left-hand side of the second Equation (6.1.14), the term $\epsilon\dot{\Phi}_i^0(0)$ is omitted, since it is small compared to other terms. Equations 6.1.14 imply that the initial data must be of the following orders: $\Phi_i(0) \sim \epsilon^2$, $\dot{\Phi}_i(0) \sim \epsilon$.

We have obtained the both terms in Equation 6.1.7 and thus found the time history of elastic displacements, provided the time history of the rigid model is given by Equations 6.1.5.

Let us calculate now the control forces and torques. The Equations 6.1.6 used above are only a part of the complete Lagrangian system of equations for our manipulators. Additional equations can be presented in the following form:

$$M_1 = -c_1\alpha_1, \qquad M_2 = -c_2\alpha_1 + c_1\beta_1 \qquad (i = 1) \quad (6.1.15)$$

$$F_1 = (m + m')(g + \ddot{z}) + m(\ddot{\beta}_2 L \cos \beta_2 + 2\dot{\beta}_2\dot{L} \cos \beta_2 - m\ddot{z} \sin^2 \beta_2)$$

$$F_2 = m\ddot{L} + m\ddot{z} \sin \beta_2 - mL\dot{\beta}_2^2, \qquad M = -c_2\alpha_2 \qquad (i = 2) \quad (6.1.16)$$

Since the functions $\alpha_i(t)$ and $\beta_i(t)$, $i = 1, 2$, are already found, whereas $z(t)$ and $L(t)$ are specified beforehand (see Equation 6.1.5), the desired control forces and torques can be readily determined from Equations 6.1.15 and 6.1.16.

Equations 6.1.4, 6.1.7, 6.1.10, and 6.1.12 determine the vector functions $\Phi_i(t) = \epsilon^2 \Phi_{i*}(t) = \|\alpha_i(t), \beta_i(t)\|$. $\dot{\Phi}_i(t)$, and $\ddot{\Phi}_i(t)$ with the accuracy up to $O(\epsilon^2)$, $O(\epsilon)$, and $O(1)$, respectively. When calculating the forces and torques using Equations 6.1.15 and 6.1.16, we retain the terms of the order of unity and drop the small (as $\epsilon \to 0$) terms. As a result, we have

$$Q_1 = \left\| \begin{matrix} M_1 \\ M_2 \end{matrix} \right\| = \left\| \begin{matrix} c_1 & 0 \\ c_2 & -c_1 \end{matrix} \right\| [C_1^{-1} B_1(t) - \epsilon^2 \theta_1(\tau)]$$

$$\tau = \epsilon^{-1} t, \qquad C_1 = O(\epsilon^{-2}) \tag{6.1.17}$$

$$Q_2 = \left\| \begin{matrix} F_1 \\ F_2 \\ M \end{matrix} \right\| = \left\| \begin{matrix} m'(\ddot{z} + g) + m(\ddot{z} + g + \ddot{\beta}_2 L) \\ m\ddot{L} \\ -c\alpha_2 \end{matrix} \right\| \tag{6.1.18}$$

In Equation 6.1.17, control torques Q_1 are expressed through the variables $B_1(t)$ and $\theta_1(\tau)$ as a sum of two terms corresponding to slow motions and fast oscillations. Equation 6.1.18 can be also reduced to a similar form:

$$Q_2 = \left\| \begin{matrix} 0 & c/L \\ 0 & 0 \\ c & 0 \end{matrix} \right\| [C_2^{-1} B_2(t) - \epsilon^2 \theta_2(t)]$$

$$- \left\| \begin{matrix} 0 & I_0/L \\ 0 & 0 \\ 0 & 0 \end{matrix} \right\| \frac{d^2 \theta_2}{d\tau^2} + \left\| \begin{matrix} m'(\ddot{z} + g) \\ m\ddot{L} \\ 0 \end{matrix} \right\| \tag{6.1.19}$$

Let us prove the equivalence of Equations 6.1.18 and 6.1.19. Using Equations 6.1.4, 6.1.10, and 6.1.7, we obtain

$$C_2^{-1} B_2 - \epsilon^2 \theta_2 = -\epsilon^2(-C_{2*}^{-1} B_2 + \theta_2) - \epsilon^2(\Phi_2^0 + \theta_2)$$

$$= -\epsilon^2 \Phi_{2*} = -\Phi_2 \tag{6.1.20}$$

From Equations 6.1.7 and 6.1.4, we have

$$\frac{d^2 \theta}{d\tau^2} = \epsilon^2 \frac{d^2}{dt^2} (\Phi_{2*} - \Phi_2^0) = \frac{d^2 \Phi_2}{dt^2} + O(\epsilon^2) \tag{6.1.21}$$

Let us insert the expressions given by Equations 6.1.20 and 6.1.21 into Equations 6.1.19, taking into account that the components of the vector Φ_2 are α_2 and β_2 (see Equation 6.1.2). Then we obtain

$$Q_2 = \left\| \begin{array}{c} -cL^{-1}\beta_2 - I_0 L^{-1}\ddot{\beta}_2 + m'(\ddot{z} + g) \\ mL \\ -c\alpha_2 \end{array} \right\| + O(\epsilon^2) \tag{6.1.22}$$

It follows from Equation 6.1.6 for $i = 2$ that the equality

$$(I_0 + mL^2)\ddot{\beta}_2 + c\beta_2 + mL(\ddot{z} + g) = 0 \tag{6.1.23}$$

is valid within the accuracy up to the terms of the order of ϵ. By solving Equation 6.1.23 for β_2 and substituting the obtained expression into Equation 6.1.22, we come from Equation 6.1.22 to Equation 6.1.18. Thus, we have proved the equivalence of Equations 6.1.18 and 6.1.19.

Equations 6.1.17 and 6.1.19 can be used for calculating control forces and torques in Problem 6.1. These forces/torques are represented as sums of two terms corresponding to slow displacements and fast oscillations of the manipulator.

6.1.3. EXAMPLES

In Chapter 5, we investigated experimentally the elastic compliance of industrial robots RPM-25 and Universal-5. It was established that the compliance is mostly conditioned by the elasticity of the joints, and the stiffness coefficients of the joints were determined. Therefore, in mechanical models of robots RPM-25 and Universal-5, we can regard the links as rigid bodies. Kinematic structures of these robots correspond to Figure 6.1.1 (RPM-25) and Figure 6.1.2 (Universal-5).

Let us apply the general results of Section 6.1.2 to these robots. Their parameters are given by

$$L_1 = 0.45 \text{ m}, \quad L_2 = 0, \quad l_1 = 0.9 \text{ m}$$

$$m_1 = 68 \text{ kg}, \quad m_2 = 154 \text{ kg}, \quad I_1 = 32 \text{ kg·m}^2, \quad I_2 = 64 \text{ kg·m}^2$$

$$c_1 = 32 \cdot 10^4 \text{ N·m}, \quad c_2 = 39 \cdot 10^4 \text{ N·m} \quad (i = 1) \tag{6.1.24}$$

$$l = 0.84 \text{ m}, \quad m = 12 \text{ kg}$$

$$I_0 = 0.48 \text{ kg·m}^2, \quad c = 2 \cdot 10^4 \text{ N·m} \quad (i = 2) \tag{6.1.25}$$

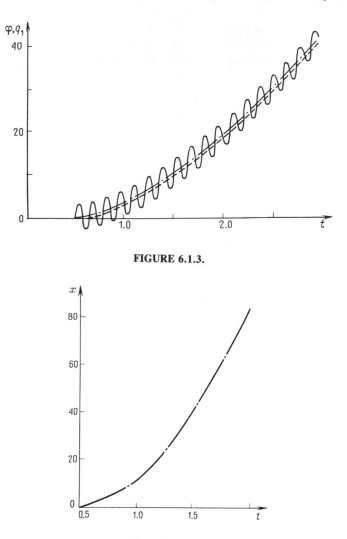

FIGURE 6.1.3.

FIGURE 6.1.4.

The data sets (Equations 6.1.24 and 6.1.25) correspond to robots RPM-25 and Universal-5, respectively. The stiffness coefficients of the joints (c_1, c_2, and c) were determined by static experiments (see Chapter 5). The equality $L_2 = 0$ in Equation 6.1.24 means that the mass center of the second link of robot RPM-25 lies on the axis of the joint O_2 (see Figure 6.1.1), i.e., this link is statically balanced with respect to its rotation axis.

The time histories (Equations 6.1.5) of the generalized coordinates were taken from experiments. In Figures 6.1.3 to 6.1.5, dashed-and-dotted lines give the variables φ (in degrees), x (in centimeters), and χ (in degrees) against time t (in seconds). We approximated the experimental curves by parabolas,

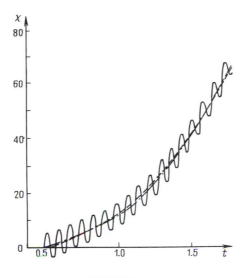

FIGURE 6.1.5.

using the least squares method with five sample points. This led to the following relationships:

$$\varphi(t) = 3.6\, t^2 + 3.7\, t - 0.8 \qquad (i = 1)$$

$$x(t) = 34.2\, t^2 - 29.1\, t$$

$$\chi(t) = 30\, t^2 - 15.6\, t - 1$$

$$L(t) = x(t) - l/2 \qquad (i = 2) \tag{6.1.26}$$

First, consider robot RPM-25 ($i = 1$). Since the second link is assumed to be statically balanced, we have $L_2 = 0$ in Equations 6.1.24. The matrix A_1 (see Equation 6.1.6) is time-independent and diagonal. We find from Equation 6.1.13 that for the arm of robot RPM-25, the eigenfrequencies of elastic oscillations are given by $\nu_{11} = \Omega_{11}/2\pi = 7$ Hz and $\nu_{12} = \Omega_{12}/2\pi = 16$ Hz. Here, $\Omega_{1k} = \epsilon^{-1}\omega_{1k}$, $k = 1, 2$. It stems from Equations 6.1.12 and 6.1.17 that the elastic displacements and control torques include rapidly oscillating components with these frequencies.

Consider a special case, where the difference between the angles φ and ψ is constant, i.e., $\delta = \varphi - \psi = \delta_0 = $ const. In this case, the links can rotate with respect to each other only by a small angle, $\beta_1(t)$. Provided $\varphi(t)$ is specified by Equation 6.1.26, we find from Equation 6.1.10 that the quasistatic angular displacements α_1^0 and β_1^0 are constant and equal to -0.3 and $-0.4°$, respectively. The time history of the generalized coordinates q_1 and q_2 (see Equations 6.1.1) that describe the motion of an anthropomorphic manipulator with elastic joints is given by

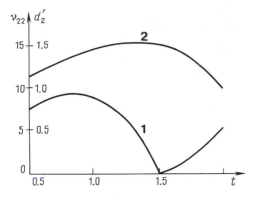

FIGURE 6.1.6.

$$\left\|\begin{matrix}q_1(t)\\q_2(t)\end{matrix}\right\| = \left\|\begin{matrix}\varphi + \alpha_1\\\psi + \beta_1\end{matrix}\right\|$$

$$= \left\|\begin{matrix}3.6\,t^2 + 3.7\,t - 0.8\\3.6\,t^2 + 3.7\,t - 0.8\end{matrix}\right\| + \left\|\begin{matrix}-0.3\\-0.4\end{matrix}\right\| + \left\|\begin{matrix}A_0\sin 14\pi t\\B_0\sin 32\pi t\end{matrix}\right\| + \left\|\begin{matrix}0\\\delta_0\end{matrix}\right\| \quad (6.1.27)$$

Here, A_0 and B_0 denote the amplitudes of fast elastic oscillations. The parameters A_0 and B_0 are determined from initial conditions.

In Figure 6.1.3, the dashed-and-dotted line presents the time history of the angle φ, the dashed curve corresponds to the sum $\varphi(t) + \alpha_1^0$, while the solid line shows the generalized coordinate q_1 vs. time. The function $q_1(t)$ is oscillating, according to Equation 6.1.27. To make the picture more comprehendable, the elastic displacements are magnified fivefold in Figure 6.1.3.

For robot Universal-5 ($i = 2$), we consider the case where the rotation of the arm and its extension are performed simultaneously. The time histories of the corresponding generalized coordinates are given by Equations 6.1.26. The motion is accompanied by the elastic oscillations with variable frequency and amplitude.

In Figure 6.1.6, the curve 1 shows the time history for the amplitude of oscillations of the mass center of the arm, $d'_{21} = L(t)d_{21}(t)$, in millimeters. The curve 2 in Figure 6.1.6 presents the frequency of these oscillations, $\nu_{21} = \Omega_{21}/(2\pi) = \omega_{21}/(2\pi\epsilon)$, in hertz, against time. The initial angular displacement d_{21}^0 (see Equations 6.1.12) is assumed to be equal to 1°. Having determined the quasistatic elastic displacements from Equation 6.1.10 and the fast elastic oscillations from Equations 6.1.12, we can find the resultant motion of the manipulator with elastic joints. In Figure 6.1.5, the dashed-and-dotted line gives the function $\chi(t)$, the dashed line presents the sum $\chi(t) + \alpha_2^0(t)$, while the solid line corresponds to the function $q(t) = \chi(t) + \alpha_2(t)$. We can see from Figure 6.1.5 that the generalized coordinate q oscillates, both the amplitude and frequency of these oscillations decreasing with time. As in Figure 6.1.3, the elastic oscillations are magnified fivefold in Figure 6.1.5.

In a similar way, we also can investigate the dynamics of the robots with elastic joints for other specified time histories of "rigid" generalized coordinates.

6.2. ALGORITHMS AND PROGRAMS FOR SIMULATING THE DYNAMICS OF ROBOTS WITH ELASTIC JOINTS

6.2.1. INTRODUCTION

In Section 6.2, we describe a software package for numerical simulation of the dynamics of multilink manipulation robots with elastically compliant joints. We suppose that the robot has the joints of the following three types: single-degree-of-freedom revolute joints, two-degrees-of-freedom revolute joints, and three-degrees-of-freedom revolute joints (spherical joints). Note that two- and three-degrees-of-freedom joints are kinematically equivalent to the combination of two or three single-degree-of-freedom joints with orthogonal axes (see Chapter 1, Section 1.1.2). Therefore, we can assume without loss of generality that the robots contain only single-degree-of-freedom revolute joints. The software package uses algorithms based on the asymptotic approach developed in Chapter 3 and allows simulation of motions of manipulators in the cases of kinematic and dynamic controls. The package contains programs that solve numerically Problems 3.1 and 3.2 formulated in Chapter 3 (Section 3.1.1) for mechanical systems with highly stiff elastic parts.

Before running the programs, it is necessary to introduce the data specifying geometrical, inertial, and elastic properties of the structure of a manipulator. One must also define Cartesian coordinate frames connected with each link of the manipulator. The ways of choosing coordinate frames and the basic kinematics of multilink manipulators are described in detail in Chapter 2.

The input data include the following parameters:

1. The number of links of a manipulator
2. The number of single-, two-, and three-degrees-of-freedom joints
3. The ordinal numbers of the joints in the kinematic chain
4. The mass of each link
5. The components of the inertia tensor of each link at its mass center, in the reference frame connected with the link
6. The coordinates of the mass center of each link, in the same reference frame
7. The direction cosines of the joint axes in the corresponding reference frames
8. The coordinates of the origin of the reference frame connected with each link, in the reference frame connected with the previous link (the

links are numbered consecutively, starting from the link connected with
the base)
9.   The stiffness coefficients of the joints.

In addition, it is necessary to specify the net force and net torque for
external forces acting on each link of the manipulator, as functions of generalized coordinates, generalized velocities, and time. As generalized coordinates, we take the angles of relative rotation of adjacent links.

The program simulating the dynamics of the robot in the case of kinematic
control requires specification of the time history of the generalized coordinates
for the "rigid" manipulator whose joints are undeformable. The program
simulating the dynamics of the robot in the case of dynamical control requires
specifying the time histories of the control torques at the joints, as well as
the initial conditions for generalized coordinates and velocities.

After the input data have been introduced, the program automatically
compiles the equations of motion and makes all computational procedures
required for solving Problems 3.1 or 3.2.

6.2.2. COMPUTER COMPILATION OF EQUATIONS OF MOTION

The approach proposed in Chapter 3 implies determining the additional
(compared with the "rigid" model) displacements, velocities, and forces
caused by the elastic compliance of the structure. The important stage of this
approach is to compile equations of the rigid model. For a multilink manipulator, the manual compilation of equations is very time consuming. The
described software package envisages automatic compiling of these equations
on the basis of the input data. Following the method of Wittenburg,[254] the
equations of motion for a rigid manipulator are derived by using the linear
and angular momenta theorems for rigid bodies. The obtained equations can
be presented as follows:

$$A(s)\ddot{s} = B(s, \dot{s}, t) + M(t) \tag{6.2.1}$$

Here, s is the vector of generalized coordinates (the relative rotation angles
for each couple of adjacent links); $A(s)$ is the matrix of the kinetic energy of
the system; $B(s, \dot{s}, t)$ is the vector-valued function including generalized inertia
forces and generalized external forces applied to the manipulator, and $M(t)$
is a vector of control torques at the joints of the robot.

According to Section 3.1.5, the solution of Problem 3.1 for a rigid manipulator includes calculating the control torques $M(t)$ from Equation 6.2.1,
provided the function $s(t)$ is specified. The solution of Problem 3.2 for a rigid
model is reduced to the numerical integration of Equation 6.2.1 under given
initial conditions and control law $M(t)$.

As mentioned in Section 3.1.5, the equations of a rigid model can have
no solution, if the rank of the matrix $A(s)$ is less than the number of degrees

of freedom. Such situation can occur for a multilink manipulator containing inertia-free links. If the rank of the matrix $A(s)$ in Equation 6.2.1 is equal to the number of degrees of freedom of the manipulator, Equations 6.2.1 have the unique solution for fixed initial data. In what follows, this condition, which can be expressed as $N \geqslant n$, is assumed to be satisfied. Here, N is the number of "inertial" coordinates, and n is the number of single-degree-of-freedom joints.

6.2.3. SIMULATION OF THE MANIPULATOR MOTION UNDER KINEMATIC CONTROL

Consider the manipulator with n single-degree-of-freedom revolute joints. Suppose the flexibility is connected with additional (compared with the rigid manipulator) rotations of the neighboring links about the axes of the joints. This assumption does not restrict the generality. If elastic rotations are possible about other axes too, we can regard these axes as the axes of additional (fictitious) joints and assume the angles of rotations about these joints to be zero for the rigid model. Denote by $q = (q_1, \ldots, q_n)$ the vector of the total angles of rotation for all pairs of adjacent links, by $s = (s_1, \ldots, s_n)$ the vector of the rotation angles for the rigid model, and by $x = (x_1, \ldots, x_n)$ the vector of the additional angles caused by the elastic compliance of the joints. This notation is the same as in Chapter 3, the equalities $N = n = m$ being held, where N, n, and m are the numbers of the "inertial", "rigid", and "elastic" coordinates, respectively. Kinematic Equation 3.1.1 in our case has the form

$$q = s + x \qquad (6.2.2)$$

For the kinematic control, we should prescribe the time history $s = s(t)$. The program computes the time histories of the coordinates x and q, as well as of the control torques M. The program whose block diagram is shown in Figure 6.2.1 solves Problem 3.1 following the methods expounded in Section 3.2.1. The calculations are carried out with a specified time step Δt. At each step, the program carries out the following operations:

1. The quasistatic component $M = M^0(t)$ of the control torque is calculated from Equation 6.2.1 for the rigid manipulator.
2. The quasistatic component y of the elastic displacement is calculated by means of Equation 3.1.28.
3. The eigenvalue problem (Equations 3.2.4 and 3.2.5) is solved, and the instantaneous frequencies $\omega_k(t)$ and the amplitude vectors $\varphi^k(t)$ are found for free elastic oscillations of the manipulator ($k = 1, \ldots, n$).
4. The oscillating component z of the elastic displacement is calculated by means of Equations 3.2.6 to 3.2.10.

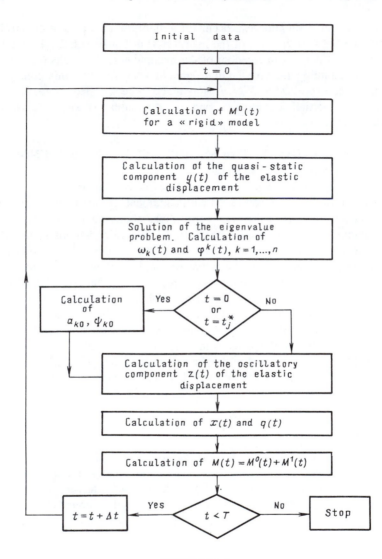

FIGURE 6.2.1.

5. The total elastic displacement x is determined according to Equation 3.2.1, and then the total rotation angles q are found from Equation 6.2.2.

6. The control torque is determined according to Equation 3.2.14 as a sum of the quasistatic and oscillating terms: $M = M^0(t) + M^1(t)$. The program stops when the current time t reaches the terminal value T.

Note that at the initial time instant $t = 0$, it is necessary to calculate the constants a_{k0} and ψ_{k0} ($k = 1, \ldots, n$) in Equations 3.2.7 and 3.2.10. These constants are found from the following equations, similar to Equations 6.1.14:

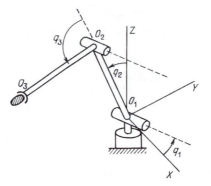

FIGURE 6.2.2.

$$\sum_{k=1}^{n} \varphi^k(0) a_{k0} \cos \psi_{k0} = \epsilon^{-2} x(0)$$

$$-\sum_{k=1}^{n} \varphi^k(0) a_{k0} \omega_k(0) \sin \psi_{k0} = \epsilon^{-1} \dot{x}(0) \qquad (6.2.3)$$

Here, $x(0)$ and $\dot{x}(0)$ are the given elastic displacement and rate of its change, and ϵ is a dimensionless small parameter (see Section 3.1.1). If some components of the vector-valued function $\ddot{s}(t) = (\ddot{s}_1(t), \ldots, \ddot{s}_n(t))$ jump at certain time instants $t = t_j^*$, $j = 1, 2, \ldots$, the quasistatic component y of the elastic displacement also jumps at these instants. Since the total elastic displacement $x(t) = y(t) + z(t)$ and velocity $\dot{x}(t) = \dot{y}(t) + \dot{z}(t)$ are continuous, we should recalculate the constants a_{k0} and ψ_{k0} at the time instants $t = t_j^*$. New values for these constants are found from the continuity conditions for the functions $x(t)$ and $\dot{x}(t)$ at the instants $t = t_j^*$. These conditions are similar to Equations 6.2.3.

Note that the integrands in Equations 3.2.7 and 3.2.10 are slowly changing functions. The characteristic time of their changes is many times as large as the periods of the natural elastic oscillations of the manipulator. Therefore, the integration can be carried out with a comparatively large step, Δt.

6.2.4. ILLUSTRATIVE EXAMPLE

Consider a two-link manipulator (Figure 6.2.2) whose links are identical homogeneous rigid bars. The dimensionless lengths and masses of the links are equal to unity. The joint O_1 connecting the first link of the manipulator with the base is a two-degrees-of-freedom joint. The joint O_2 connecting the links is a single-degree-of-freedom revolute joint, its axis being normal to the manipulator plane. At the end O_3 of the second link, a gripper with a load is placed. The dimensions of the gripper and the load are small compared with the link lengths, and the gripper together with the load is regarded as a unit point mass. The control torques M_1, M_2, and M_3 are applied at the axes of the joints. No external forces, apart from the reaction of the base, act upon the system.

The joint O_1 includes two axes of rotation, one of them is fixed in space, whereas the other is perpendicular to the first one.

Let us introduce the fixed coordinate frame O_1XYZ whose origin is placed at the joint O_1, and the axis O_1Z is pointed along the fixed axis of the joint.

Denote by q_1 the angle between the axis O_1X and the second (moving) axis of the joint O_1; by q_2, the angle between the link O_1O_2 and the axis O_1Z; by q_3, the angle between the links; by s_i ($i = 1, 2, 3$), the angles q_i in the absence of deformations in the joints (the generalized coordinates of the "rigid" model); by x_i ($i = 1, 2, 3$), the additional (due to the elastic compliance) angles of rotation about the joint axes. The variables $q = (q_1, q_2, q_3)$, $s = (s_1, s_2, s_3)$, and $x = (x_1, x_2, x_3)$ are related by Equation 6.2.2. The angles q_1, q_2, and q_3 are measured as shown in Figure 6.2.2. All these angles are equal to zero, if the moving axis of the joint O_1 is directed along O_1X and the links lie on the axis O_1Z.

For simulation, we choose $s(t)$ such that the gripper O_3 moves along the broken line $K_0K_1K_2$ shown in Figure 4.5.1 (see Section 4.5.3). This line lies in the plane $x = -0.45$, the coordinates of the vertices K_0, K_1, and K_2 in the reference frame O_1XYZ are given by

$$K_0 = (-0.45, 0.84, 0.52)$$
$$K_1 = (-0.45, 0.33, 0.82)$$
$$K_2 = (-0.45, 0.33, 1.42)$$

On each segment, K_0K_1 and K_1K_2, the time history of the motion of the gripper, is the same. At first, the gripper accelerates for the time $\tau_1 = 0.17$, then moves with a constant velocity for the time $\tau_2 = 0.71$, decelerates for the time $\tau_3 = \tau_1 = 0.17$, and stops.

The total time of motion from the point K_0 to K_2 is $T = 4\tau_1 + 2\tau_2 = 2.1$. The constant velocity on the middle stages is $v = 0.68$; on the acceleration and deceleration stages, the velocity changes linearly between 0 and v. Thus, we have specified the time histories of the projections r_x^0, r_y^0, and r_z^0 of the radius-vector $\mathbf{r}^0 = \overrightarrow{O_1O_3}$ of the gripper onto the axes of the reference frame O_1XYZ. The functions $s_i(t)$ are given by Equations 4.4.6, where one should replace α_i by s_i and put $l = 1$. Note that each position of the gripper matches two manipulator configurations which differ by the sign of the angle s_3 (or q_3). When calculating, we have chosen the configuration corresponding to the positive angle s_3 (the sign " $+$ " in the last expression in Equation 4.4.6.

We specify the elastic potential energy by

$$U = \frac{1}{2} k(x_1^2 + x_2^2 + x_3^2), \qquad k = 10^4 \qquad (6.2.4)$$

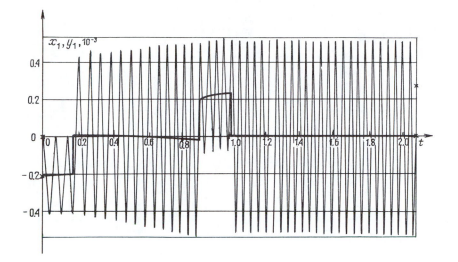

FIGURE 6.2.3.

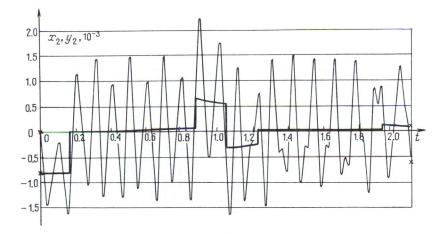

FIGURE 6.2.4.

According to Equation 3.1.6, the small parameter ϵ is of the order of 0.01. Therefore, the asymptotic approach is applicable here.

The results of simulating the manipulator motion are given in Figures 6.2.3 to 6.2.8. Figures 6.2.3 to 6.2.5 show the time histories for the quasistatic (y_i) and total (x_i) elastic displacements. The time histories of the quasistatic components (M_i^0) and the total values (M_i) of the control torques are presented in Figures 6.2.6 to 6.2.8. The solid, thick curves show the quasistatic components. Qualitatively, the motion of the manipulator is close to that discussed in Section 4.5.3 for the manipulator with elastic links.

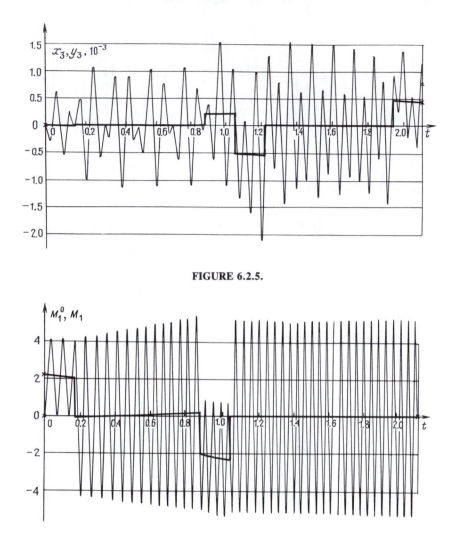

FIGURE 6.2.5.

FIGURE 6.2.6.

6.2.5. SIMULATION IN THE CASE OF DYNAMIC CONTROL

The program simulating the motion of a manipulator under the dynamic control solves Problem 3.2 according to the method expounded in Section 3.2.3. We consider only those systems for which the number N of the "inertial" generalized coordinates $q = (q_1, \ldots q_N)$ is equal to the number n of the generalized coordinates $s = (s_1, \ldots, s_n)$ of the "rigid" model. The number m of additional "elastic" generalized coordinates $x = (x_1, \ldots, x_m)$ can be arbitrary.

In the case of dynamic control, we are to prescribe the time history of the control torques $M(t)$, as well as the initial conditions for the variables s

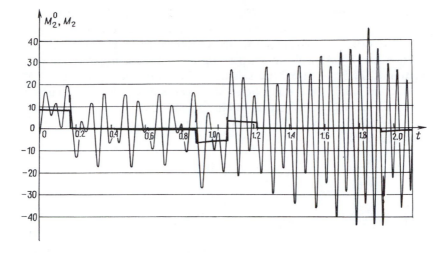

FIGURE 6.2.7.

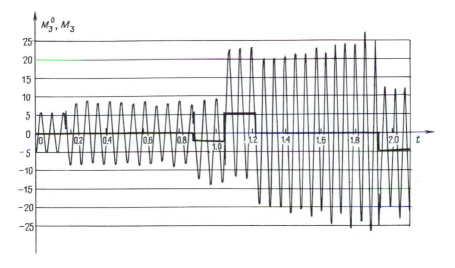

FIGURE 6.2.8.

and $\dot{s}$. The program calculates the time histories of the variables s, x, and q, thus determining the motion of a manipulator with elastic joints. The program integrates numerically the equations of motion for an auxiliary fictitious rigid manipulator (see Section 3.2.3). These equations are represented in the form similar to Equation 6.2.1:

$$A(\sigma)\ddot{\sigma} = B(\sigma, \dot{\sigma}, t) + M(t) + \Delta M(\sigma, t)$$

Here, the additional torques $\Delta M(\sigma, \text{t})$ are calculated at each integration step in accordance with the technique of Section 3.2.3 (see Equations 3.2.33 and 3.2.34 for Q^{σ}). Initial conditions for the variable σ are determined according to Equations 3.2.33 and 3.2.36. The variables s, x, and q that are to be determined in Problem 3.2 are calculated by means of Equations 3.2.32, 3.2.33, and 3.2.37. The integration and all calculations are carried out with a certain time step, Δt. As mentioned in Chapter 3, the step Δt can be rather large, as compared with the periods of natural elastic oscillations of the manipulator.

Chapter 7

OPTIMAL CONTROL OF MANIPULATION ROBOTS

7.1. OPTIMAL CONTROL IN ROBOTICS (A SURVEY)

As mentioned in Chapter 1, a manipulation robot is a multipurpose machine able to carry out various motions. The change of motion does not require alterations in the mechanical structure of the robot and is performed by changing the control program only. When planning a technological process involving a robot, one is to choose the motion of the manipulator for accomplishing a prescribed operation. A natural reasonable approach to this choice is to optimize the motion of the manipulator with respect to some performance criterion that depends on the parameters of the robot, the technological task, and environment. The optimal motion is provided by the appropriate control law. Thus, we come to optimal control problems for robots (see Section 2.1.3). The solution of these problems is based on theory and methods of optimal control and mathematical programming.

The number of scientific publications on optimization in robotics is large and grows rapidly. Several directions of research can be distinguished in this field; namely, optimization of motion along a prescribed trajectory, optimization of trajectories of manipulators, optimization of kinematically redundant manipulators, and optimization of control for flexible robots. In what follows, we give a brief survey of these directions.

7.1.1. OPTIMIZATION OF ROBOT MOTION WITH A PRESCRIBED TRAJECTORY OF THE GRIPPER

Here, the problem is stated as follows: find the control law that drives the gripper of a manipulator along a prescribed trajectory from a given initial position to a given terminal one and minimizes a given performance index. This problem is natural for technological operations that require the gripper to move along a prescribed trajectory with a definite orientation at each point. Such requirements are typical for welding, cutting, inspecting, and other operations.

The basic idea of the majority of publications on the above problem is to reduce it to the optimal control problem for a single-degree-of-freedom system. The trajectory parameter (e.g., the length of its arc measured from the initial position of the gripper), as a rule, is considered as a sole generalized coordinate.

In References 32,33, and 84, the time-optimal control algorithm for a multilink, nonredundant manipulator is developed. The control variables are the torques created by the drives at the manipulator joints. These torques are subject to constraints, depending on the state variables. The initial problem is reduced to the optimal control problem for a simple mechanical system equivalent to the point mass driven along the straight line by the control force. However, constraints on the state and control variables can be rather complex. It is proved,[33] under certain assumptions, that at each time instant the acceleration of the gripper assumes either maximal or minimal admissible values. Based on this result, the algorithm for determining the switching points of the optimal control is suggested. This algorithm is modified in Reference 209 for the case, where the optimal motion contains singular arcs, on which the acceleration of the gripper lies between the upper and lower bounds. Similar methods are developed also in References 190 and 226. Another technique for treating the same problem is used in Reference 188, where the trajectory parameter is regarded as an independent variable and the geometry of admissible domains for squared velocities and their derivatives with respect to the trajectory parameter is considered.

The described approach is widespread in scientific publications on robotics. In References 31,81,82,119,194, and 223, the algorithm proposed in References 32, 33, and 84 is used for more complicated optimization problems, in particular, for trajectory optimization under collision avoidance conditions (see References 31,82,223).

In some papers, trajectories of special kind and performance indices other than time are considered.

For example, Luh and Lin[157] deal with time-optimal motions of a robot whose gripper moves along a trajectory consisting of straight segments and circular arcs. In Reference 227, instead of a continuous trajectory, certain points are specified which the robot must pass through during its motion.

The combined integral functional, taking into account the time, the averaged kinetic energy, and the driving torques, is considered by Pfeiffer and Johanni.[190]

7.1.2. OPTIMIZATION OF MANIPULATOR TRAJECTORIES

This direction is connected with the conventional optimal control problem stated as follows: find the control driving a manipulator from a given initial state (or manifold) to a given target state (or manifold) and to minimize the prescribed cost functional (performance index). In the general case, both the state and control variables are constrained. Here, the trajectory is not specified beforehand, but is determined as a result of optimization. Below, we discuss three approaches to the optimization of robot trajectories, namely, the direct use of optimal control methods, parametric optimization, and the separation of motions.

7.1.2.1. Direct Use of Optimal Control Methods

This way is most efficient for systems with a small number of degrees of freedom.

General mathematical theorems establishing the existence and structure of optimal control in different robotic motions are proved in References 2,49,50, and 215.

Nosov et al.[179,180] consider manipulators described by a system of decoupled linear differential equations of the first and second order (a Cartesian-coordinate manipulator can serve as an example). They obtain the explicit solution for optimal control problems with different performance indices and constraints.

A number of publications are devoted to the time-optimal control of a two-link planar manipulator.[2,7,39–41,67,92,163,181,249,250,253] Ailon and Langholz[2] proved the existence of optimal control for the point-to-point motion of a two-link manipulator, provided the control torques are restricted. In References 92,163,181,249,250, and 253, time-optimal motions for a two-link manipulator are calculated numerically by methods based on Pontryagin's maximum principle. Note that in References 163,181,249, and 250 the optimal control ensuring the point-to-point motion of the gripper is found together with the optimal placement of the robot.

Osipov and Formalskii[184] consider cylindrical- and spherical-coordinate robots controlled by restricted forces and torques. The possibility of the time-optimal control law with an infinite number of switches is established. In Reference 46, such chattering controls are rigorously studied.

A considerable number of publications deal with numerical calculation of optimal motions of manipulation robots. These publications contain the description of computational algorithms used for solving optimal control problems (see, e.g., References 155,199,242), as well as numerous examples of optimal trajectories for robots of various kinematic structures.[92,114,139–141,217]

The majority of authors study minimum-time motions of manipulation robots. Some optimal control problems regarding energy consumption for the motion of a manipulator are considered in References 16,17,134,218, and 219.

7.1.2.2. Parametric Optimization

In this approach, the initial admissible set of controls is replaced by a more narrow one, namely, by a parametric family with a comparatively small number of parameters. Hence, the optimal control problem is reduced to minimization of a function of several variables. This approach is used quite often because it simplifies the computational procedure substantially. Note that the parametric optimization, strictly speaking, is not a method for solution of the original optimal control problem, but the replacement of this problem by another one. The success of this approach depends on the choice of a parametric family of controls (or trajectories) to be optimized.

It follows from the maximum principle that, if the equations of the system are linear in control variables, the nonsingular time-optimal control is of the bang-bang type. That is why some authors seek optimal control in the class of bang-bang functions, the switch instants being regarded as the parameters to be found (see, e.g., References 19,67,135,136,160,161,240).

Ritz's method is used in References 156 and 207. The optimal trajectory is presented as a linear combination of some basis functions with unknown coefficients. The coefficients are determined from the optimality conditions. In Reference 51, cubic splines are used to approximate the trajectory.

A number of authors use the parametric optimization combined with the method[33] related to a prescribed trajectory. In this approach, the parametric family of trajectories connecting the initial and terminal positions of a manipulator is specified. For each sample of the parameter vector (i.e., for each trajectory), the optimal control can be obtained. By using the mathematical programming technique, the parameters are determined that correspond to the minimal value of the performance criterion. This method is employed in References 31,81–83, and 194.

7.1.2.3. Separation of Motions

In this approach, the original nonlinear system of coupled equations is replaced by a more simple system consisting of independent subsystems. The separation of motions can be achieved by an appropriate change of variable in the equations of motion, by ignoring small terms, by using asymptotic perturbation methods, by an appropriate choice of control, and also by a combination of these techniques.

Kahn and Roth[125] consider the time-optimal control problem for a three-link anthropomorphic manipulator. They linearize the equations of motion in the vicinity of the terminal state and then, by using an appropriate change of variables, reduce the initial problem to three independent optimal control problems for single-degree-of-freedom systems. After that, optimal control is obtained analytically in the feedback form.

In References 63,65, and 66, a method of control is proposed for a general Lagrangian system with controlled degrees of freedom. This system is reduced to the set of linear subsystems, the nonlinear and coupling terms being regarded as disturbances. If the level of the disturbances for each degree of freedom is less than the maximal admissible value of the corresponding control, the feedback control law is obtained that brings the original system to the terminal state in finite time. The control is based on the game theory, it is time-suboptimal, and robust. If the disturbances due to the dynamical interaction of different degrees of freedom are small, they are often ignored when constructing the control. In Reference 65, the limiting level for the disturbances is established that still guarantees the arrival of the system at the desired state for such control laws.

The methods for decoupling the dynamic equations of robots by means of appropriate control laws that compensate the dynamic interaction of different degrees of freedom are expounded in many papers and books (e.g., see References 87,146,147, and 224).

To conclude this section, we mention some publications on optimal trajectory planning for manipulation robots of the most-used designs. Different approaches (analytical and numerical) to time-optimal control for a two-link anthropomorphic manipulator are discussed by Chernousko et al.[67] Both open-loop and feedback controls are obtained. In a series of papers,[139–141,217] time-optimal trajectories and open-loop controls are constructed and investigated numerically for an anthropomorphic manipulator,[140] a polar-coordinate manipulator,[139,141] and a SCARA-type robot.[217] In Reference 217, optimal motions of the anthropomorphic and SCARA-type robots are compared. Time-optimal and near-optimal controls for a cylindrical-coordinate manipulator with irreversible gear trains are constructed by Akulenko et al.[8] Berbyuk and Yanchak[30] calculated time-optimal transport motions of a gantry robot. Newman[173] deals with the optimal control of manipulators designed so that the dynamic interaction between different links is eliminated. Takano and Susaki[223] and Zhang and Wang[262] constructed optimal trajectories of manipulation robots in the presence of obstacles in their working zones.

The papers in References 6,16–18,36,37,94,202–204, and 206 are devoted to optimal control of manipulation robots with electric drives. The voltages applied to the motors are regarded as control variables. In most of these papers, the optimality criterion is the time of motion, while in References 16,17, and 204 the energy consumption is included into the performance index.

7.1.3. OPTIMIZATION OF REDUNDANT MANIPULATORS

If the number of degrees of freedom of a manipulator exceeds that of the manipulation object, the manipulator is called kinematically redundant. For such robots, an infinite number of manipulator configurations match each position of the manipulation object, and many trajectories in the configuration space of the manipulator correspond to each trajectory of the gripper in the physical space. Kinematic redundancy makes a robot more versatile and allows it to avoid collisions with obstacles by the choice of appropriate configurations. It also implies the possibility of optimizing manipulator motion, provided that the motion of the gripper is prescribed.

In the majority of publications, the optimization problem for redundant manipulators is stated as follows: given a prescribed motion of the gripper of the robot, find the motion of the manipulator so as to minimize either a scalar function of the state variables at each time instant (the local optimization) or a functional that depends on the motion as a whole (the global optimization). The local optimization is considered, for example, in References 113,118,131,153,243,251,252, and 255. As performance criteria, the sum of squared generalized velocities of the manipulator[255] and its kinetic energy[251,252] are used. Sometimes, other criteria are employed; for instance, the total driving power.[131] The global optimization is treated, for example, in References 132,138,133, and 170. Time integrals of the sum of squared generalized velocities or the kinetic energy are usually taken as the performance indices.

Time-optimal motions of kinematically redundant manipulators are considered in References 7 and 127 under the assumption that each degree of freedom moves independently, according to the prescribed law. This assumption, verified experimentally for certain types of robots, entails that the motion of the robot as a whole is determined, if the initial and terminal configurations of the manipulator are specified. Optimal configurations and motions of kinematically redundant robots are obtained by means of the special numerical algorithm.[7,127] These results were applied to the welding robot.[102]

7.1.4. OPTIMAL CONTROL OF ROBOTS WITH ELASTIC PARTS

The elasticity of links, joints, and gear trains of a manipulator causes oscillations, thus increasing operational time and reducing accuracy. Therefore, when designing control for flexible manipulators, we are to envisage an efficient suppression of elastic oscillations. Optimal control for such manipulators can be found by the special methods of optimal control of oscillatory systems.[3,68]

In References 73,191,197,198, and 260, near time-optimal open-loop controls are constructed for a single-link electromechanical manipulator whose reduction gear possesses an elastic compliance. In References 73 and 198, the bang-bang control of the arm of the rotating robot is obtained for specified initial and terminal positions. Rogov[197] deals with the control that stops the motion of the arm and quenches the elastic oscillations. Papers in References 191 and 260 are devoted to parametric optimization.

In a number of publications, optimal and suboptimal control laws for robot arms with distributed elasticity are constructed. Akulenko and Mikhailov[11] obtain an approximate time-optimal feedback control for a single-link electromechanical manipulator, taking into account the elastic compliance of both the arm and the reduction gear. The optimal open-loop control for a rotating elastic link is found in Reference 9. The cost functional is the time integral of squared angular acceleration.

In the book by Berbyuk[25] and in Berbyuk et al.,[26–29] different methods are used for optimization of motions of manipulators with elastic links. Parametric optimization, combined with inverse methods of dynamics, is employed as well as the approach based on the first integrals of motion. A number of results on optimal control and optimal design of flexible robots are presented.

For a more detailed survey on the optimization of control laws and structural parameters of manipulation robots see Bolotnik and Chernousko.[34]

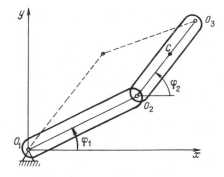

FIGURE 7.2.1.

7.2. TIME-OPTIMAL CONTROL OF A TWO-LINK MANIPULATOR

7.2.1. STATEMENT OF THE PROBLEM

In Sections 7.2 and 7.3, we consider time-optimal control problems for a two-link manipulator regarding the joint torques as control variables. A two-member linkage is a part of many manipulation robots of an anthropomorphic design. The manipulator of such a robot consists of an upper arm, a forearm, and a hand with a gripper. The upper arm and forearm are usually bars connected with each other by means of revolute joints, with the upper arm being attached to the base. To provide high maneuverability of the manipulator, the hand is made of a number of links whose linear dimensions are generally much less than those of the upper arm and forearm. The upper arm and forearm perform transport motions which bring the hand to the desired position. The prescribed orientation of a payload in the gripper is implemented by rotating the links of the hand. Transport motions, as a rule, take more time than orientational ones. Hence, optimization of transport motions with respect to operational time seems to be of most importance from the practical point of view. In the majority of practical cases, the linear dimensions of a payload are small compared with those of the manipulator. Hence, we can treat the hand with a payload as a mass point and neglect orientational motions. Therefore, when considering time-optimal transport motions it is often sufficient to use a two-link model of a manipulator.

Consider a two-link manipulator (Figure 7.2.1) consisting of two rigid links, G_1 and G_2, connected by the joint, O_2. The first link, G_1, is connected with a fixed base by means of the joint O_1. Both joints are revolute, with their axes being parallel to each other. At the end (O_3) of the second link (G_2), a gripper is placed which holds a payload regarded as a point mass.

The described mechanical system performs planar motions, with the trajectories of all points being parallel to the plane Π which passes through the point O_3 and is perpendicular to the axes of the joints O_1 and O_2. In what follows, we denote by the same letters the joints themselves and the points of their intersection with the plane Π.

The manipulator is controlled by means of two drives, D_1 and D_2, located at the points O_1 and O_2, respectively. Net torques (with respect to the axes of the joints O_1 and O_2) created by the drives D_1 and D_2 and applied to the links G_1 and G_2, respectively, are denoted by M_1 and M_2. The action of other forces, apart from the reaction of the base, is not taken into account.

Let us introduce in the plane Π a fixed reference frame, O_1xy, with the origin at the point O_1. Denote by φ_1 the angle between the axis O_1x and the straight line O_1O_2 connecting the joints; by φ_2 the angle between the axis O_1x and the straight line O_2C passing through the projection C of the mass center of the second link (together with the gripper and load) onto the plane Π; by x and y the Cartesian coordinates of the load; by $L = |O_1O_2|$ the distance between the axes of the joints (the length of the first link); by $L_2 = |O_2O_3|$ the distance between the axis of the joint O_2 and the gripper (the length of the second link); by $L = |O_2C|$ the distance between the axis of the joint O_2 and the mass center of the link G_2 with the load; by I_1 and I_2 the moments of inertia of the links G_1 and G_2 (with the load) about the axes of the joints O_1 and O_2, respectively; by m_2 the mass of the second link with the load; and by m the mass of the load.

Let us take the angles φ_1 and φ_2 as the generalized coordinates. Then the kinetic energy of the manipulator is given by

$$T = \frac{1}{2}(I_1 + m_2L_1^2)\dot{\varphi}_1^2 + \frac{1}{2}I_2\dot{\varphi}_2^2 + m_2L_1L\cos(\varphi_1 - \varphi_2)\dot{\varphi}_1\dot{\varphi}_2 \quad (7.2.1)$$

The virtual work of control forces is expressed as

$$\delta A = Q_1\delta\varphi_1 + Q_2\delta\varphi_2, \qquad Q_1 = M_1 - M_2, \qquad Q_2 = M_2 \quad (7.2.2)$$

In accordance with Equations 7.2.1 and 7.2.2, the Lagrange Equations 2.3.11 for our mechanical system take the form

$$(I_1 + m_2L_1^2)\ddot{\varphi}_1 + m_2L_1L\cos(\varphi_1 - \varphi_2)\ddot{\varphi}_2$$

$$+ m_2L_1L\sin(\varphi_1 - \varphi_2)\dot{\varphi}_2^2 = M_1 - M_2$$

$$I_2\ddot{\varphi}_2 + m_2L_1L\cos(\varphi_1 - \varphi_2)\ddot{\varphi}_1 - m_2L_1L\sin(\varphi_1 - \varphi_2)\dot{\varphi}_2^2 = M_2 \quad (7.2.3)$$

The Cartesian coordinates x and y of the load are expressed through the generalized coordinates as follows:

$$x = L_1\cos\varphi_1 + L_2\cos\varphi_2, \quad y = L_1\sin\varphi_1 + L_2\sin\varphi_2 \quad (7.2.4)$$

The angles φ_1 and φ_2 are defined through the Cartesian coordinates not uniquely. By solving Equations 7.2.4 for φ_1 and φ_2 in the domain $-\pi \leq \varphi_1 \leq \pi$ ($i = 1, 2$), we obtain

$$\varphi_i = \Phi_i - \pi(1 + \text{sign}(|\Phi_i| - \pi))\text{sign}(\Phi_i - \pi), \quad i = 1, 2$$

$$\Phi_1 = \tan^{-1}\frac{y}{x} - \Gamma\delta_1 + \frac{\pi}{2}(1 - \text{sign } x)\,\text{sign } y$$

$$\Phi_2 = \tan^{-1}\frac{y}{x} + \Gamma\delta_2 + \frac{\pi}{2}(1 - \text{sign } x)\,\text{sign } y$$

$$\delta_1 = \cos^{-1}\frac{x^2 + y^2 + L_1^2 - L_2^2}{2L_1\sqrt{x^2 + y^2}}, \quad \delta_2 = \cos^{-1}\frac{x^2 + y^2 + L_2^2 - L_1^2}{2L_2\sqrt{x^2 + y^2}}$$

$$x^2 + y^2 \leq (L_1 + L_2)^2, \quad \Gamma = \text{sign } \Psi = \pm 1, \quad \Psi = \varphi_2 - \varphi_1 \qquad (7.2.5)$$

Equations 7.2.5 reflect the fact that each position of the load inside the working zone $[x^2 + y^2 < (L_1 + L_2)^2]$ corresponds to two configurations of the manipulator with different signs Γ of the angle between the links, $\Psi = \varphi_2 - \varphi_1$. The configuration with $\Gamma = 1$ ($0 < \Psi < \pi$) is shown by a solid line in Figure 7.2.1, and the configuration with $\Gamma = -1$ ($-\pi < \Psi < 0$) is drawn by a dashed line. The variables δ_1 and δ_2 in Equation 7.2.5 are the angles at the vertices O_1 and O_3 (respectively) of the triangle $O_1O_2O_3$.

Problem 7.1. Find the time histories of the control torques $M_1(t)$ and $M_2(t)$, transferring the system described by Equation 7.2.3 from the initial state,

$$\varphi_1(0) = \varphi_1^0, \quad \dot{\varphi}_1(0) = 0, \quad \varphi_2(0) = \varphi_2^0, \quad \dot{\varphi}_2(0) = 0 \quad (7.2.6)$$

to the specified terminal state,

$$\varphi_1(T) = \varphi_1^1, \quad \dot{\varphi}_1(T) = 0, \quad \varphi_2(T) = \varphi_2^1, \quad \dot{\varphi}_2(T) = 0 \quad (7.2.7)$$

in the minimal time T, provided the absolute values of the control torques are constrained by

$$|M_1(t)| \leq M_{10}, \quad |M_2(t)| \leq M_{20} \qquad (7.2.8)$$

If the type of the manipulator configuration (i.e., the parameter Γ) is specified at the beginning and the end of the control process, Problem 7.1 can be reformulated in terms of Cartesian coordinates of the load as follows.

Problem 7.2. Find the time histories of the control torques $M_1(t)$ and $M_2(t)$, transferring the load from the initial state,

$$x(0) = x_0, \qquad y(0) = y_0, \qquad \dot{x}(0) = \dot{y}(0) = 0$$

$$\Gamma(0) = \Gamma_0, \qquad x_0^2 + y_0^2 \le (L_1 + L_2)^2, \qquad \Gamma_0 = \pm 1 \qquad (7.2.9)$$

to the specified terminal state,

$$x(T) = x_1, \qquad y(T) = y_1, \qquad \dot{x}(T) = \dot{y}(T) = 0$$

$$\Gamma(T) = \Gamma_1, \qquad x_1^2 + y_1^2 \le (L_1 + L_2)^2, \qquad \Gamma_1 = \pm 1 \qquad (7.2.10)$$

in the minimal time T, provided the control torques are constrained by Equation 7.2.8.

The initial and terminal values of the Cartesian and angular coordinates in Equations 7.2.6, 7.2.7, 7.2.9, and 7.2.10 are related by Equations 7.2.4 and 7.2.5.

7.2.2. INERTIA-FREE MANIPULATOR

Suppose the mass of the manipulator is much less than the mass of the load, and that both links have the same length l. The first assumption is fulfilled for special robots designed for manipulating large payloads in space. As for the second assumption, it is usually satisfied. Note that for the given total length of the links ($2l$), the manipulator with equal links has the maximal working zone, namely, a circle $x^2 + y^2 \le 4l^2$.

We consider that the links of the manipulator are inertia-free and the mass of the whole system is concentrated at the point O_3 and is equal to the mass of the load. Hence, we put $L_1 = L_2 = L = l$, $I_1 = 0$, $I_2 = ml^2$, and $m_2 = m$ in Equations 7.2.1 to 7.2.5.

Let us take the Cartesian coordinates of the load, x and y, as the generalized coordinates. Then the kinetic energy is given by

$$T = \frac{1}{2} m(\dot{x}^2 + \dot{y})^2 \qquad (7.2.11)$$

The generalized forces Q_x and Q_y, corresponding to the generalized coordinates x and y, respectively, are expressed in terms of the generalized forces Q_1 and Q_2 (see Equation 7.2.2) as follows:

$$Q_x = Q_1 \frac{\partial \varphi_1}{\partial x} + Q_2 \frac{\partial \varphi_2}{\partial x}, \qquad Q_y = Q_1 \frac{\partial \varphi_1}{\partial y} + Q_1 \frac{\partial \varphi_2}{\partial y} \qquad (7.2.12)$$

Substituting Equations 7.2.2 and 7.2.5 into Equation 7.2.12, we obtain

$$Q_x = (M_1 - 2M_2)R\Gamma x - M_1 y\rho^{-2}$$

$$Q_y = (M_1 - 2M_2)R\Gamma y + M_1 x\rho^{-2}$$

$$R = \rho^{-1}(4l^2 - \rho^2)^{-1/2}, \qquad \rho = (x^2 + y^2)^{1/2} \qquad (7.2.13)$$

The Lagrange Equations 2.3.11 compiled in accordance with Equations 7.2.11 and 7.2.13 have the form

$$m\ddot{x} = \frac{(M_1 - 2M_2)\Gamma x}{\sqrt{(x^2 + y^2)(4l^2 - x^2 - y^2)}} - \frac{M_1 y}{x^2 + y^2}$$

$$m\ddot{y} = \frac{(M_1 - 2M_2)\Gamma y}{\sqrt{(x^2 + y^2)(4l^2 - x^2 - y^2)}} + \frac{M_1 x}{x^2 + y^2} \qquad (7.2.14)$$

The geometry of a two-link manipulator (see Figure 7.2.1) implies that the type of its configuration (i.e., the parameter Γ) does not change during the motion, if the trajectory of the load does not touch the circle $x^2 + y^2 = 4l^2$ (which is the boundary of the manipulator working area) and does not pass through the origin ($x = 0$, $y = 0$). If $x^2 + y^2 = 4l^2$ or $x = y = 0$, the right-hand sides of Equations 7.2.14 have singularities. The singularity on the boundary of the working zone ($x^2 + y^2 = 4l^2$) is connected with the possibility of a shock, while the degeneration at the origin ($x = y = 0$) is due to the vanishing of the moment of inertia of the system about the axis of the joint O_1.

We will solve Problem 7.2 for Equations 7.2.14, assuming that the type of the manipulator configuration does not change during the motion [$\Gamma(t) \equiv \Gamma_0 = \Gamma_1$] and the control torques M_1 and M_2 are restricted by the same constant: $M_{10} = M_{20}$ in Equations 7.2.8. We also suppose that $y_0 = y_1 = y_*$ in Equations 7.2.9 and 7.2.10. This condition does not restrict generality and can be provided by rotating the reference frame by the angle

$$\alpha = \tan^{-1}[(y_1 - y_0)/(x_1 - x_0)]$$

Let us introduce new (dimensionless) variables:

$$x' = x/l, \qquad y' = y/l, \qquad t' = l^{-1}(M_{10}m)^{1/2}t \qquad (7.2.15)$$

In what follows, the primes marking the dimensionless variables are omitted. In these variables, the equations of motion, initial conditions, terminal conditions, and constraints are given by Equations 7.2.14 and 7.2.8 to 7.2.10, with $M_{10} = M_{20} = 1$, $L_1 = L_2 = l = 1$, and $m = 1$.

To describe all possible optimal motions, we are to solve Problem 7.2 for each type of configuration ($\Gamma = 1$ and $\Gamma = -1$) and for all values of the parameters x_0, x_1, and y_* satisfying the inequalities

$$|y_*| \leq 2, \qquad |x_0| \leq \sqrt{4 - y_*^2}, \qquad |x_1| \leq \sqrt{4 - y_*^2} \qquad (7.2.16)$$

Equations 7.2.16 in the three-dimensional (x_0, x_1, y_*)-space is the intersection of two circular cylinders of the radius $\rho = 2$ whose axes coincide with the coordinate axes Ox_0 and Ox_1. The intersection lines of these cylinders are ellipses in the planes $x_0 + x_1 = 0$ and $x_0 - x_1 = 0$. The domain described by Equations 7.2.16 is symmetric with respect to the planes $x_0 + x_1 = 0$ and $x_0 - x_1 = 0$. Intersections of this domain with the planes $y_* = $ const are squares whose projections onto the coordinate plane Ox_0x_1 define the admissible domain of the parameters x_0 and x_1 for fixed y_*. A direct substitution shows that equations of motion (Equations 7.2.14) are invariant with respect to the following transformations:

$$x \rightarrow -x, \qquad y \rightarrow -y \qquad\qquad (7.2.17)$$

$$x \rightarrow -x, \qquad \Gamma \rightarrow -\Gamma, \qquad M_i \rightarrow -M_i, \qquad i = 1, 2 \quad (7.2.18)$$

$$t \rightarrow T - t \qquad\qquad (7.2.19)$$

Let $\{M_1(t), M_2(t), x(t), y(t), T\}$ be the solution of the time-optimal control problem for Equations 7.2.14 with the boundary conditions (Equations 7.2.9 and 7.2.10) where $y_0 = y_1 = y_*$. The invariance with respect to Equations 7.2.17 to 7.2.19 entails the following statements.

Statement 7.1. With the change $x_0 \rightarrow -x_0$, $x_1 \rightarrow -x_1$, $y_* \rightarrow -y_*$, $\Gamma \rightarrow \Gamma$, the solution of the time-optimal control problem becomes $\{M_1(t), M_2(t), -x(t), -y(t), T\}$.

Statement 7.2. With the change $x_0 \rightarrow -x_0$, $x_1 \rightarrow -x_1$, $y_* \rightarrow y_*$, $\Gamma \rightarrow -\Gamma$, the solution of the time-optimal control problem becomes $\{-M_1(t), -M_2(t), -x(t), y(t), T\}$.

Statement 7.3. With the change $x_0 \rightarrow x_1$, $x_1 \rightarrow x_0$, $y_* \rightarrow y_*$, $\Gamma \rightarrow \Gamma$, the solution of the time-optimal control problem becomes $\{M_1(T - t), M_2(T - t) x(T - t), y(T - t), T\}$.

If the solutions of Problem 7.2 for $\Gamma = 1$ and $\Gamma = -1$ are known for all values of parameters x_0, x_1, and y_* belonging to one octant of the domain specified by Equations 7.2.16, e.g., for $y_* \geqslant 0$, $x_0 \geqslant x_1$, $x_0 \geqslant -x_1$, then the solutions of this problem for the rest admissible (satisfying Equations 7.2.16) values of parameters are readily obtained by means of Statements 7.1 to 7.3 Thus, the above mentioned invariance leads to a significant reduction in computations.

7.2.3. NUMERICAL SOLUTION

In the general case, the optimal control problem for the nonlinear Equation 7.2.14 does not admit an exact analytical solution. We solved Problem 7.2 numerically, using the method of successive approximations[71] based on Pontryagin's maximum principle.[192] Optimal trajectories of the payload under the

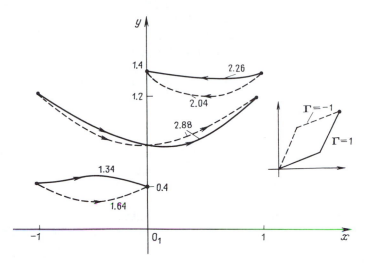

FIGURE 7.2.2.

initial conditions (Equation 7.2.9) and terminal conditions (Equation 7.2.10) are shown in Figure 7.2.2 for a number of values of the parameters x_0, x_1, and $y_* > 0$ ($y_* = y_0 = y_1$). The trajectories corresponding to $\Gamma = 1$ and $\Gamma = -1$ are given by solid and dashed lines, respectively. The arrows indicate the direction of motion. The numbers above the curves are the dimensionless times of motions along the respective trajectories. We can see from Figure 7.2.2 that the optimal trajectories and minimal times depend on the type of the manipulator configuration. Note that, if $x_0 = -x_1$, the optimal times coincide for both types of configuration. This stems from Statements 7.2 and 7.3. These statements also imply that for $x_0 = -x_1$, the optimal trajectories corresponding to different types of configuration of the manipulator are related by

$$x_-(t) = -x_+(T - t), \qquad y_-(t) = y_+(T - t) \qquad (7.2.20)$$

Here and below, the indices $+$ and $-$ mark the variables corresponding to $\Gamma = 1$ and $\Gamma = -1$, respectively. It follows from Equations 7.2.20 that for $x_0 = -x_1$ the optimal trajectories corresponding to $\Gamma = 1$ and $\Gamma = -1$ can be obtained from each other by reflecting with respect to the axis O_1y.

The calculations show that it is possible to obtain a considerable gain in operational time by choosing an appropriate type of manipulator configuration. Table 7.2.1 contains the optimal times, T_+ (for $\Gamma = 1$) and T_- (for $\Gamma = -1$), for $x_0 = 1$, $x_1 = 0$, and for different $y_* > 0$. The last row of the table gives the relative difference η of the optimal times corresponding to different configurations of the manipulator; $\eta = |T_+ - T_-|/\min(T_+, T_-)$. Note that η grows if y_* decreases. For small y_*, the magnitude of η can be

TABLE 7.2.1
Optimal Times of Motion Depending on the
Configuration of the Manipulator

y_*	0.4	0.6	0.8	1.0	1.2	1.4	1.6
T_+	1.34	1.48	1.61	1.74	1.89	2.04	2.37
T_-	1.64	1.74	1.87	1.96	2.16	2.26	2.39
η	0.224	0.176	1.161	0.126	0.143	0.108	0.008

substantial; for example, we have $\eta = 0.224$ when $y_* = 0.4$. Therefore, in some cases, it is advisable to choose the optimal type of manipulator configuration when planning transport motions.

7.2.4. OPTIMAL RECTILINEAR MOTIONS

Here, we consider controls that transfer the load from the state given by Equation 7.2.9 to the state in Equation 7.2.10 along the straight line connecting the initial and terminal positions. As in Sections 7.2.2 and 7.2.3, we suppose that the control torques M_1 and M_2 are restricted by the same constant M_0 ($|M_i| \leq M_0$, $i = 1, 2$) and the type of manipulator configuration does not change during the motion. Without loss of generality, we assume $y_0 = y_1 > 0$ in Equations 7.2.9 and 7.2.10 (see also Section 7.2.2). Then the y coordinate of the load is constant for a rectilinear motion: $y(t) \equiv y_0$. We describe the motion of the system governed by Equations 7.2.14 using dimensionless variables (Equations 7.2.15); therefore, the control constraints are given by

$$|M_1| \leq 1, \qquad |M_2| \leq 1 \qquad (7.2.21)$$

We substitute $y(t) = y_0$ into the second Equation 7.2.14 and express M_2 from this equation:

$$M_2 = M_1[1 + \Gamma f(x, y_0)]/2$$

$$f(x, y_0) = xy_0^{-1}[4(x^2 + y_0^2)^{-1} - 1]^{1/2}$$

$$(x, y_0) \in D = \{x, y_0: x^2 + y_0^2 \leq 4\} \qquad (7.2.22)$$

Inserting M_2 from Equation 7.2.22 and $y = y_0$ into the first Equation 7.2.14, we obtain the differential equation for the variable x:

$$\ddot{x} = u, \qquad u = -M_1/y_0 \qquad (7.2.23)$$

It follows from Equations 7.2.21 and 7.2.22 that the control u in Equation 7.2.23 must satisfy the constraint

$$|u| \leq \mu(x) \qquad (7.2.24)$$

$$\mu(x) = \min\left\{\frac{1}{y_0}, \ \frac{2}{y_0|1 + \Gamma f(x, y_0)|}\right\} > 0 \qquad (7.2.25)$$

Note that the function $\mu(x)$ depends on the parameters y_0 and Γ, which are constant for each trajectory.

Thus, the time-optimal control problem for a rectilinear motion of the load is defined by Equations 7.2.23 and 7.2.24 and the boundary conditions

$$x(0) = x_0, \qquad \dot{x}(0) = 0, \qquad x(T) = x_1, \qquad \dot{x}(T) = 0 \quad (7.2.26)$$

Let us assume, without loss of generality, that $x_0 \le x_1$. In case $x_0 > x_1$, we can make the following change of variables: $x \rightarrow -x$, $u \rightarrow -u$. We consider only such motions in which the coordinate x changes monotonically and, therefore, $\dot{x} \ge 0$. Then Equations 7.2.23 and 7.2.24 imply the inequalities

$$-\mu(x)\dot{x} \le \ddot{x}\dot{x} \le \mu(x)\dot{x}$$

$$(7.2.27)$$

Integration of the right Inequality 7.2.27 from 0 to t under the initial conditions specified by Equations 7.2.26 gives

$$\dot{x} \le \Psi_0(x), \qquad \Psi_0(x) = \left[2\int_{x_0}^{x} \mu(\xi)d\xi\right]^{1/2} \qquad (7.2.28)$$

It stems from here that the time t in which the coordinate of the load can reach the value $x \ge x_0$ satisfies the relationship

$$t \ge \int_{x_0}^{x} \frac{d\xi}{\Psi_0(\xi)} \qquad (x \ge x_0) \qquad (7.2.29)$$

The equality signs in Inequalities 7.2.28 and 7.2.29 correspond to the control $u = \mu(x)$ and the phase trajectory γ_0 satisfying the equation

$$\dot{x} = \Psi_0(x) \qquad (7.2.30)$$

Analogously to Equation 7.2.29, the time $T - t$, in which the coordinate of the load can change from some value $x < x_1$ to x_1 and meet the terminal condition specified in Equations 7.2.26, satisfies the inequality

$$T - t \ge \int_{x}^{x_1} \frac{d\xi}{\Psi_1(\xi)}, \qquad \Psi_1(x) = \left[2\int_{x}^{x_1} \mu(\xi)d\xi\right]^{1/2} \qquad (7.2.31)$$

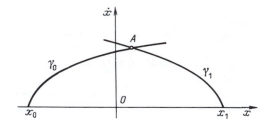

FIGURE 7.2.3.

The equality sign in Equation 7.2.31 takes place on the phase trajectory γ_1, corresponding to the control $u = -\mu(x)$ and satisfying the equation,

$$\dot{x} = \Psi_1(x) \qquad (7.2.32)$$

similar to Equation 7.2.30. The phase trajectories γ_0 and γ_1 are shown in Figure 7.2.3.

Equations 7.2.28 and 7.2.31 for the functions $\Psi_0(x)$ and $\Psi_1(x)$ imply that $\Psi_0(x)$ monotonically increases, whereas $\Psi_1(x)$ monotonically decreases with x. Hence, the curves γ_0 and γ_1 intersect at the point $A = (x_A, \dot{x}_A)$ on the phase plane. It follows from Inequalities 7.2.29 and 7.2.31 that the time of the motion T satisfies the inequality

$$T \geq \int_{x_0}^{x_A} \frac{d\xi}{\Psi_0(\xi)} + \int_{x_A}^{x_1} \frac{d\xi}{\Psi_1(\xi)} \qquad (7.2.33)$$

The equality in Equation 7.2.33 is attained on the phase trajectory consisting of the arc of the curve γ_0 [from the point $(x_0, 0)$ to A] and the arc of the curve γ_1 [from A to the terminal $(x_1, 0)$]; see Figure 7.2.3. This trajectory is the only optimal trajectory corresponding to the minimal time of motion of the manipulator. The coordinates x_A and $\dot{x}_A$ of the intersection point A can be found from the equations

$$\dot{x}_A = \Psi_0(x_A) = \Psi_1(x_A) \qquad (7.2.34)$$

The instant τ of passing the point A is the instant of switch of the control. The instant τ and the optimal time T are given by

$$\tau = \int_{x_0}^{x_A} \frac{d\xi}{\Psi_0(\xi)}, \qquad T = \tau + \int_{x_A}^{x_1} \frac{d\xi}{\Psi_1(\xi)} \qquad (7.2.35)$$

The optimal control is unique and expressed by

$$u(t) = \mu(x(t)), \qquad t \in (0, \tau)$$
$$u(t) = -\mu(x(t)), \qquad t \in (\tau, T) \qquad (7.2.36)$$

Using Equations 7.2.36, 7.2.22, and 7.2.23 we find the optimal time histories for the control torques M_1 and M_2:

$$M_1(t) = \mu(x(t))y_0\text{sign}[(\tau - t)(x_0 - x_1)]$$
$$M_2(t) = M_1(t)[1 + \Gamma f(x, y_0)]/2 \qquad (7.2.37)$$

Equations 7.2.37 are applicable to both cases $x_0 \leq x_1$ and $x_0 > x_1$.

Under the control given by Equations 7.2.37, the load moves from the initial position towards the terminal position with acceleration on the time interval $0 < t < \tau$, and with deceleration on the time interval $\tau \leq t < T$. At each time instant $t \in [0, T]$, one of the control torques (M_1 or M_2) assumes its maximal admissible absolute value, i.e.,

$$|M_1| = 1, \quad \text{if} \quad \mu(x) = 1/y_0$$
$$|M_2| = 1, \quad \text{if} \quad \mu(x) = 2y_0^{-1}|1 + \Gamma f(x, y_0)|^{-1}$$

These formulas follow from Equations 7.2.25 and 7.2.37.

Equation 7.2.25 admits two possibilities. If

$$|1 + \Gamma f(x, y_0)| \leq 2 \qquad (7.2.38)$$

for all x belonging to the segment connecting the initial and terminal positions of the load, then $\mu(x) = y_0^{-1} = \text{const}$. In this case, the solution of the time-optimal control problem (Equations 7.2.23 to 7.2.26) is well known[47,192] and given by

$$M_1(t) = \text{sign}[(\tau - t)(x_0 - x_1)] \qquad (7.2.39)$$

The corresponding minimal time is equal to

$$T_0 = 2(|x_1 - x_0|y_0)^{1/2} \qquad (7.2.40)$$

Since T_0 is the minimal time of motion under the constraint $|M_1| \leq 1$, the time T from Equation 7.2.35 satisfies the inequality $T \geq T_0$ for all admissible motions. In Inequality 7.2.38, we have $T = T_0$, whereas in the opposite case $T > T_0$.

Let us determine the domains $G(\Gamma)$, $\Gamma = \pm 1$, in the (x, y_0) plane, where Inequality 7.2.38 holds together with $y_0 \geq 0$ and $x^2 + y^2 \leq 4$. The last

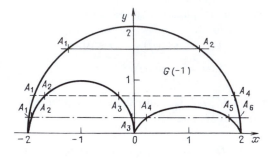

FIGURE 7.2.4.

inequality defines the working zone of the manipulator with equal links in the dimensionless variables (see Equations 7.2.5 and 7.2.15 with $L_1 = L_2$). We obtain

$$G(-1) = \{x, y_0: [1 - (1 + x)^2]^{1/2} \le y_0 \le (4 - x^2)^{1/2}, \quad -2 \le x < 0;$$

$$\frac{1}{3}[2x(4x^2 + 9)^{1/2} - 5x^2]^{1/2} \le y_0 \le (4 - x^2)^{1/2}, \qquad 0 \le x \le 2\}$$

$$G(1) = \{x, y_0: \frac{1}{3}[-2x(4x^2 + 9)^{1/2} - 5x^2]^{1/2} \le y_0 \le (4 - x^2)^{1/2},$$

$$-2 \le x < 0;$$

$$[1 - (1 - x^2)]^{1/2} \le y_0 \le (4 - x^2)^{1/2}, \qquad 0 \le x \le 2\} \tag{7.2.41}$$

The domain $G(-1)$ is bounded by the half circle of the radius 2 with the center at the origin, the half circle of the unity radius with the center at the point $(x = -1, y_0 = 0)$, and the curve

$$y_0 = \chi(x) = 1/3[2x(4x^2 + 9)^{1/2} - 5x^2]^{1/2}, \qquad 0 \le x \le 2$$

At the point $x = (3/4)^{1/2}$ the function $\chi(x)$ reaches its maximum, equal to 0.5. The domain $G(-1)$ is shown in Figure 7.2.4, whereas the domain $G(1)$ can be obtained by reflecting $G(-1)$ with respect to the axis $O_1 y_0$.

If the manipulator configuration is $\Gamma = -1$ ($\Gamma = 1$) and the segment connecting the initial (x_0, y_0) and terminal (x_1, y_0) points belongs to the domain $G(-1)$ [or $G(1)$], the optimal time is given by Equations 7.2.40.

7.2.5. OPTIMAL CONFIGURATIONS

We have shown in Section 7.2.3 that the optimal time of motion depends substantially on the type of manipulator configuration. In this connection let us consider the following problem.

Problem 7.3. Find the optimal type of manipulator configuration (i.e., the optimal value of the parameter $\Gamma = \pm 1$) corresponding to the minimal time possible for moving the load along a rectilinear trajectory with $y = y_0 = \text{const}$ between the boundary states (Equations 7.2.26) under the optimal control (Equations 7.2.37).

Problem 7.3 should be solved in the domain

$$P(y_0) = \{x_0, x_1: |x_0| \leq (4 - y_0^2)^{1/2}, \qquad |x_1| \leq (4 - y_0^2)^{1/2}\} \quad (7.2.42)$$

for each $y_0 \in (0, 2)$. Equations 7.2.42 follow from Equations 7.2.16. The solution of Problem 7.3 implies a partition of the domain given by Equations 7.2.42 into three subdomains. In two of them, the optimal value $\Gamma(x_0, x_1, y_0)$ is equal to 1 and -1, respectively, whereas in the third subdomain both values $\Gamma = \pm 1$ correspond to the same optimal time.

Let us determine these subdomains.

First, we find the domains $\Pi(y_0, \Gamma)$ corresponding to the optimal transfer for the time T_0 (see Equation 7.2.40). On the (x, y_0) plane, we draw a line $y = y_0$ ($0 < y_0 < 2$) which intersects the boundary of the domain $G(\Gamma)$ at several points. The number K of intersection points depends on y_0. Denote these points and their abscissae, numbered in the increasing order, by A_i and $x^{(i)}$, respectively, $i = 1, \ldots, K; K \leq 6$; see Figure 7.2.4. The segments A_1A_2, A_3A_4, and A_5A_6 shown in Figure 7.2.4 belong to the domain $G(\Gamma)$. The union of the squares

$$\Pi_i = \{x_0, x_1: x^{(i)} \leq x_0 \leq x^{(i+1)}, \qquad x^{(i)} \leq x_1 \leq x^{(i+1)}\} \quad (7.2.43)$$

corresponding to these segments is the sought-for domain $\Pi(y_0, \Gamma)$.

Following the above procedure and considering different cases, we arrive at the following results. For $y_0 \geq 1$, the domain $\Pi(y_0, \Gamma)$ is the entire square $P(y_0)$ (see Equations 7.2.42), for both $\Gamma = 1$ and $\Gamma = -1$. If $0.5 \leq y_0 < 1$, then the domain $\Pi(y_0, \Gamma)$ consists of two squares of the type presented in Equations 7.2.43. If $0 < y_0 < 0.5$, then $\Pi(y_0, \Gamma)$ consists of three squares of the type presented by Equations 7.2.43. For example, the domains $\Pi(0.8, -1)$ and $\Pi(0.4, -1)$ are shaded in Figure 7.2.5, a and b, respectively. Since the sets $G(-1)$ and $G(1)$ from Equations 7.2.41 transform into each other with the change of x for $(-x)$, the domains $\Pi(y_0, 1)$ and $\Pi(y_0, -1)$ convert into each other under the central symmetry transformation of the plane Ox_0x_1.

The definition of the domains $\Pi(y_0, \Gamma)$ implies that for all $(x_0, x_1) \in \Pi(y_0, -1) \cap \Pi(y_0, 1)$, the optimal time is independent of the configuration Γ and equal to T_0.

For the pairs (x_0, x_1) such that

$$(x_0, x_1) \in P(y_0) \backslash [\Pi(y_0, -1) \cup \Pi(y_0, 1)]$$

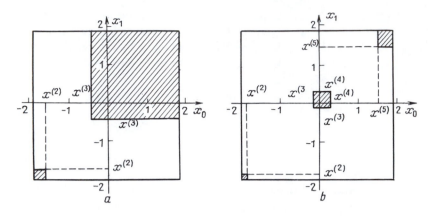

FIGURE 7.2.5.

Problem 7.3 was solved numerically. For each type Γ of the manipulator configuration, the time of the transfer T was calculated according to Equations 7.2.35, and then the optimal configuration $\Gamma(x_0, x_1, y_0)$ was chosen that corresponded to the least T. It follows from Statements 7.1 to 7.3 given in Section 7.2.2 that the function $\Gamma(x_0, x_1, y_0)$ satisfies the following relationships:

$$\Gamma(x_0, x_1, y_0) = \Gamma(-x_0, -x_1, -y_0)$$

$$\Gamma(x_0, x_1, y_0) = -\Gamma(-x_0, -x_1, y_0)$$

$$\Gamma(x_0, x_1, y_0) = \Gamma(x_1, x_0, y_0)$$

Figure 7.2.6 shows the partition of the square $P(y_0)$ from Equation 7.2.42 generated by the function $\Gamma(x_0, x_1, y_0)$ for $y_0 = 0.7$. A qualitative picture of the partition is similar for other $y_0 < 1$ too. The regions corresponding to optimal configurations with $\Gamma = -1$ and $\Gamma = 1$ are shaded by lines with a different slope. The double hatching shows the domain $\Pi(y_0, -1) \cap \Pi(y_0, 1)$ where the control (Equation 7.2.37) provides the same time for transferring the load, irrespective of the type of manipulator configuration Γ. The boundary lines between different regions presented in Figure 7.2.6 also possess such a property.

7.2.6. COMPARISON OF CONTROL LAWS

Let us compare the minimal times required for transferring the load along the optimal (see Section 7.2.3) and rectilinear (see Sections 7.2.4 and 7.2.5) trajectories. We also estimate the gain in time due to the optimal choice of manipulator configuration Γ in the case of rectilinear trajectories.

Denote by $T^0(x_0, x_1, y_0)$ the time of transferring the load along the optimal trajectory from the state specified by Equation 7.2.9 to the state specified by

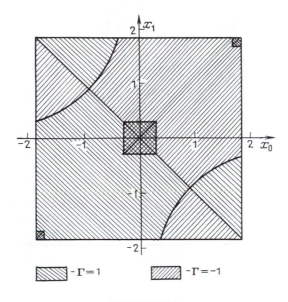

FIGURE 7.2.6.

Equation 7.2.10, if the type of configuration is chosen in the best way. To determine T^0 for fixed x_0, x_1, y_0, it is necessary to solve the optimal control problems for Equations 7.2.14 with $\Gamma = 1$ and $\Gamma = -1$ (see Section 7.2.2), and then to choose the least time of motion. Denote by $T(x_0, x_1, y_0, \Gamma)$ the time of transferring the load from the state specified by Equation 7.2.9 to the state specified by Equation 7.2.10 along a rectilinear trajectory with the use of the control given by Equations 7.2.37.

Let us introduce the following quantities:

$$\delta(x_0, x_1, y_0) = \frac{T(x_0, x_1, y_0, \Gamma^0) - T^0(x_0, x_1, y_0)}{T^0(x_0, x_1, y_0)}$$

$$\nu(x_0, x_1, y_0) = \frac{T(x_0, x_1, y_0, -\Gamma^0) - T(x_0, x_1, y_0, \Gamma_0)}{T(x_0, x_1, y_0, \Gamma_0)}$$

where $\Gamma^0 = \Gamma^0(x_0, x_1, y_0) = \pm 1$ is the optimal configuration of the manipulator transferring the load along a rectilinear trajectory in accordance with the control given by Equations 7.2.37. The quantity δ is a relative difference of the times of transfer corresponding to the optimal control and the control law given by Equations 7.2.37, while ν characterizes the gain due to the optimal choice of Γ in rectilinear motions.

Figure 7.2.7 presents δ (solid line) and ν (dashed line) against the parameter y_0 for fixed $x_0 = -1$, $x_1 = 0$. The graph of the function $\nu(y_0)$ shows that for the control laws given by Equation 7.2.37 the time of motion sub-

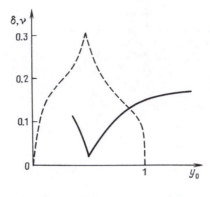

FIGURE 7.2.7.

stantially depends on the configuration of the manipulator. The maximal value of ν is attained at $y_0 = 0.5$ and is equal to 0.32. The quantity δ does not exceed 0.17 for $0.4 \leq y_0 \leq 1.4$.

Note that the numerical calculation of the optimal control, i.e., solution of Problem 7.2 for Equations 7.2.14, is more difficult and time consuming than the determination of the optimal rectilinear motion. Therefore, since δ is relatively small, it seems practically advisable to use simple rectilinear trajectories instead of the optimal ones, if a technological process does not require the productivity of the robot to be as high as possible.

7.3. PARAMETRIC OPTIMIZATION: A SEMI-INVERSE APPROACH

7.3.1. PRELIMINARY REMARKS

In Section 7.3 we consider Problem 7.1 for a two-link manipulator with arbitrary geometrical and inertial parameters. The motion of the manipulator is governed by differential Equations 7.2.3. Let us introduce dimensionless variables and parameters,

$$t' = \left(\frac{M_{20}}{m_2 L_1^2}\right)^{1/2} t, \qquad L' = \frac{L}{L_1}$$

$$I_i' = \frac{I_i}{m_2 L_1^2}, \qquad M_i' = \frac{M_i}{M_{20}}, \qquad i = 1, 2 \qquad (7.3.1)$$

and substitute them into Equations 7.2.3, 7.2.6, 7.2.7, and 7.2.8. If we omit primes and turn the reference frame $O_1 xy$ by the angle φ_1^0, we arrive at the same Equations 7.2.3 and 7.2.6 through 7.2.8 with

$$\varphi_1^0 = 0, \qquad m_2 = 1, \qquad L_1 = 1, \qquad M_{20} = 1 \qquad (7.3.2)$$

The explicit solution of Problem 7.1 can be found only for some particular cases, e.g., for the case where the center of mass of the second link coincides with the point O_2 (see Section 7.3.3). In the general case, a numerical solution can be obtained. However, it takes a lot of computer time and cannot cover all possible values of the parameters I_1, I_2, L, M_{10}, φ_2^0, φ_1^1, and φ_2^1. Therefore, it is desirable to develop a simple and efficient approach for constructing and presenting optimal or near-optimal controls.

7.3.2. SEMI-INVERSE APPROACH

It follows from Pontryagin's maximum principle that nonsingular time-optimal controls for the system governed by Equations 7.2.3 with 7.2.8 are of the bang-bang type. We consider controls of the simplest structure in which one of the control torques, M_1 or M_2, has only one switch, while the other has two switches during the motion. These control laws can be presented in one of the following forms

$$(a)\ M_1 = M_{10}\mu_\alpha(t), \qquad M_2 = \nu_\beta(t)$$

$$(a)\ M_1 = M_{10}\nu_\beta(t), \qquad M_2 = \mu_\alpha(t)$$

$$\mu_\alpha(t) = (-1)^\alpha \mathrm{sign}(\tau_0 - t)$$

$$\nu_\beta(t) = \begin{cases} (-1)^\beta, & \text{if}\quad t \in [0, \tau_1) \cup [\tau_2, T] \\ (-1)^{\beta+1}, & \text{if}\quad t \in [\tau_1, \tau_2) \end{cases}$$

$$0 \le \tau_0 \le T, \qquad 0 \le \tau_1 \le \tau_2 \le T, \qquad \alpha, \beta = 0, 1 \qquad (7.3.3)$$

Both torques $M_1(t)$ and $M_2(t)$ in Equations 7.3.3 satisfy Equations 7.2.8 and assume maximal admissible absolute values. Equations 7.3.3 imply eight possible types of control laws which differ from each other by the number of the torque with one switch (variants [a] and [b]) and by the order of sign alternation for each torque (determined by combinations of values α, β = 0, 1). Let us mark these types of control programs by the number k = 1, 2, . . . , 8 in the following way:

$$k = 1: \ (a)\ \alpha = 0,\ \beta = 0 \qquad k = 5: \ (a)\ \alpha = 0,\ \beta = 1$$

$$k = 2: \ (a)\ \alpha = 1,\ \beta = 1 \qquad k = 6: \ (a)\ \alpha = 1,\ \beta = 0$$

$$k = 3: \ (b)\ \alpha = 0,\ \beta = 0 \qquad k = 7: \ (b)\ \alpha = 0,\ \beta = 1$$

$$k = 4: \ (b)\ \alpha = 1,\ \beta = 1 \qquad k = 8: \ (b)\ \alpha = 1,\ \beta = 0 \qquad (7.3.4)$$

The control of each type has the minimal number (four) of free parameters (τ_0, t_1, τ_2, and T) to be determined from four terminal conditions (Equations 7.2.7)

$$\varphi_i^{(k)} (T; \tau_0, \tau_1, \tau_2) = \varphi_i^1, \qquad i = 1, 2 \qquad (7.3.5)$$

$$\dot{\varphi}_i^{(k)} \left(T; \tau_0, \tau_1, \tau_2 \right) = 0, \qquad i = 1, 2 \qquad (7.3.6)$$

Here, $\varphi_i^{(k)} \left(t; \tau_0, \tau_1, \tau_2 \right)$ is the solution of Equations 7.2.3 with the initial conditions (Equations 7.2.6) for the control of the type k (see Equations 7.3.4); τ_0, τ_1, and τ_2 are the switching times, and T is the terminal time. It is possible to find the parameters τ_0, τ_1, τ_2, and T using conventional shooting numerical proceedings. These require many iterations and include multiple integration of Equations 7.2.3. In addition, the convergence of methods for solving Equations 7.3.5 and 7.3.6 is problematic. For this reason, we suggest a simple semi-inverse procedure to construct special diagrams for calculating the parameters of the control laws. For a fixed initial value $\varphi_2 = \varphi_2^0$, the procedure consists of the following stages for each type k of the control programs (Equations 7.3.4).

1. Fix two parameters τ_1 and τ_2 and find numerically the remaining parameters τ_0 and T from the velocity Equations 7.3.6. The time T is determined in the course of numerical integration of Equations 7.2.3 from one of the Equations 7.3.6. The parameter τ_0 is found by shooting from the second Equation 7.3.6. Thus, we reduce our problem with four free parameters to the search for only one parameter.
2. Using the coordinate Equations 7.3.5, determine the terminal position $(\varphi_1^1, \varphi_2^1)$ corresponding to the parameters τ_0, τ_1, τ_2, and T just found.
3. Carry out stages 1 and 2 for a set of values of τ_1 and τ_2, changing them by certain small steps. Thus we obtain the set Ω_k of terminal positions $(\varphi_1^1, \varphi_2^1)$ reachable from the fixed initial state by means of controls of type k.
4. Plot the lines of equal values of the parameters τ_0, τ_1, τ_2, and T inside the set Ω_k.

An example of the set Ω_k for $k = 1$, $I_1 = 1$, $I_2 = 1$, $L = 0.1$, $M_{10} = 2$, and $\varphi_2^0 = 0.5$ rad is shown in Figure 7.3.1. The typical time histories of the torques M_1 and M_2 for $k = 1$ are also shown in the figure.

To obtain the complete diagram, we must find all the domains Ω_k, $k = 1, 2, \ldots, 8$ which can overlap. Hence, we are to choose the type k which provides the least time T.

To reduce the amount of calculations, we restrict ourselves to Equations 7.3.4 with $k \leq 4$. To some extent, this restriction is justified by the fact that for the case $L = 0$, optimal controls can be always taken with $k \leq 4$ (see Section 7.3.3).

Figure 7.3.2 presents a typical diagram corresponding to $I_1 = 1$, $I_2 = 1$, $L = 0.1$, $M_{10} = 2$ and $\varphi_2^0 = 0.5$ rad. The sets Ω_k marked by the numbers $k = 1, 2, 3, 4$ do not overlap and cover all the plane φ_1^1, φ_2^1. The thick lines show the boundaries of Ω_k. Lines of $\tau_1 = $ const and $\tau_* = \tau_2 - 2\tau_1 = $ const are plotted as solid curves. The dashed and dotted curves correspond to lines

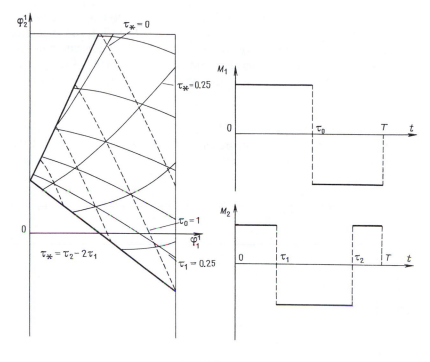

FIGURE 7.3.1.

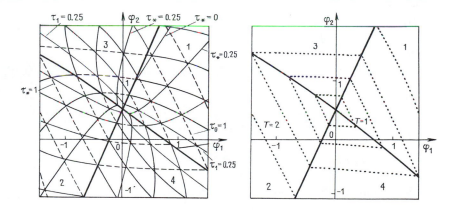

FIGURE 7.3.2.

of $\tau_0 = $ const and $T = $ const, respectively. Diagrams similar to Figure 7.3.2 are obtained for different values of the parameters I_1, I_2, L, M_{10}, and φ_2^0. Using these diagrams, we can determine the type k of the control, which depends on the set Ω_k containing the given terminal point (φ_1^1, φ_2^1). Furthermore, using lines of equal values we can obtain times of switches τ_0, τ_1, and τ_2 as well as the terminal time T.

Note that in the cases where the explicit solution of Problem 7.1 is obtained (for $L = 0$ and $|\varphi_1^1 - \varphi_1^0| \ll 1$, $|\varphi_2^1 - \varphi_2^0| \ll 1$, see Section 7.3.3), the control programs determined by means of the diagram technique coincide with the optimal controls. Very likely, this also holds for a wider range of the parameters. In general, control laws calculated by means of the diagrams described above can be regarded as near optimal.

7.3.3. EXACT OPTIMAL SOLUTIONS

Let us solve Problem 7.1 for the case where the mass center C of the second link lies on the axis of the joint O_2. In this case $L = 0$, and Equations 7.2.3 become

$$(I_1 + 1)\ddot{\varphi}_1 = M_1 - M_2, \qquad I_2\ddot{\varphi}_2 = M_2 \qquad (7.3.7)$$

Here, we take Equations 7.3.2 into account. The change of variables

$$\psi_1 = (I_1 + 1)\varphi_1 + I_2\varphi_2, \qquad \psi_2 = I_2\varphi_2 \qquad (7.3.8)$$

reduces Equations 7.3.7 to the form

$$\ddot{\psi}_1 = M_1, \qquad \ddot{\psi}_2 = M_2 \qquad (7.3.9)$$

According to Equations 7.2.6, 7.2.7, 7.3.2, and 7.3.8, the boundary conditions are given by

$$\psi_1(0) = \psi_1^0 = I_2\varphi_2^0, \qquad \dot{\psi}_1(0) = 0,$$

$$\psi_2(0) = \psi_2^0 = I_2\varphi_2^0, \qquad \dot{\psi}_2(0) = 0 \qquad (7.3.10)$$

$$\psi_1(T) = \psi_1^1 = (I_1 + 1)\varphi_1^1 + I_2\varphi_2^1, \qquad \dot{\psi}_1(T) = 0$$

$$\psi_2(T) = \psi_2^1 = I_2\varphi_2^1, \qquad \dot{\psi}_2(T) = 0 \qquad (7.3.11)$$

The solution of the time-optimal control problem for each Equation 7.3.9 is well known and expressed as follows:

$$M_i^* = M_{i0}\text{sign}[(T_i/2 - t)\Delta\psi_i], \qquad T_i = 2(|\Delta\psi_i|/M_{i0})^{1/2} \qquad (7.3.12)$$

$$\Delta\psi_i = \psi_i^1 - \psi_i^0, \qquad i = 1, 2$$

The control $M_i^*(t)$ transfers the system $\ddot{\psi}_i = M_i$, subject to constraint $|M_i| \leq M_{i0}$ from the state $\psi_i(0) = \psi_i^0$, $\dot{\psi}_i(0) = 0$ to the state $\psi_i(T_i) = \psi_i^1$, $\dot{\psi}_i(T_i) = 0$ in the minimal time $T = T_i$.

Using Equations 7.3.12, we can readily construct the time-optimal control for the system described by Equations 7.3.9 to 7.3.11. At first, we calculate T_1 and T_2 according to Equations 7.3.12 and note that the minimal time is given by $T = \max(T_1, T_2)$.

If $T_1 \geq T_2$, then the control $M_1(t)$ is determined uniquely, and $M_1(t) = M_*^1(t)$. The control $M_2(t)$ is not unique and can be chosen in the following bang-bang form,

$$M_2(t) = \nu_\beta(t), \qquad \beta = 0, 1 \qquad (7.3.13)$$

with not more than two switches. Here, the function $\nu_\beta(t)$ is defined in Equations 7.3.3.

Let us determine the instants of switches τ_1 and τ_2 from the condition of bringing the coordinate ψ_2 to the terminal state at the time $T = T_1$. Integrating the second Equation 7.3.9 under the initial conditions given by Equations 7.3.10 and the control $M_2(t)$ from Equation 7.3.13, we get

$$\psi_2(T_1) = \psi_2^0 + (-1)^\beta[\tau_2^2 - \tau_1^2 - 2T_1(\tau_2 - \tau_1) + T_1^2/2]$$

$$\dot{\psi}_2(T_1) = (-1)^\beta[T_1 - 2(\tau_2 - \tau_1)], \qquad \beta = 0, 1 \qquad (7.3.14)$$

Inserting $\psi_2(T_1)$ and $\dot{\psi}_2(T_1)$ from (Equation 7.3.14) into the terminal conditions (Equation 7.3.11), we obtain two equations for τ_1 and τ_2. Solving these equations, we find

$$\tau_1 = (-1)^\beta \Delta\psi_2/T_1 + T_1/4, \qquad \tau_2 = (-1)^\beta \Delta\psi_2/T_1 + 3T_1/4$$

$$T_1 = 2(|\Delta\psi_1|/M_{10})^{1/2}, \qquad \beta = 0, 1 \qquad (7.3.15)$$

A direct comparison shows that $0 \leq \tau_1 \leq \tau_2 \leq T_1$ in the domain $|\Delta\psi_2| \leq |\Delta\psi_1|/M_{10}$ corresponding to $T_1 \geq T_2$.

The controls $M_1 = M_*^1(t)$ and $M_2 = \nu_\beta(t)$, given by Equations 7.3.12 and 7.3.13, with the parameters τ_1 and τ_2 determined by Equations 7.3.15 are optimal in the case $T_1 \geq T_2$ and transfer Equation 7.3.9 from the state specified by Equations 7.3.10 to the state specified by Equations 7.3.11 in the minimal time $T = T_1$. Note that both values $\beta = 0$ and $\beta = 1$ in Equation 7.3.13 correspond to optimal controls.

If $T_1 \leq T_2$, the optimal controls M_1 and M_2 are given by

$$M_1 = M_{10}\nu_\beta(t), \qquad \tau_1 = (-1)^\beta \Delta\psi_1/(M_{10}T_2) + T_2/4$$

$$\tau_2 = (-1)^\beta \Delta\psi_1/(M_{10}T_2) + 3T_2/4$$

$$M_2 = \text{sign}[(T_2/2 - t)\Delta\psi_2] \qquad (7.3.16)$$

and the optimal time $T = T_2$. If $T_1 = T_2 = T$, the optimal control is unique and given by Equations 7.3.12. In this case, Equations 7.3.13, 7.3.15, and 7.3.16 coincide with Equation 7.3.12, and each torque has only one switch. Equations 7.3.12, 7.3.13, 7.3.15, and 7.3.16 give the solution to Problem 7.1 for the case L = 0 in terms of variables ψ_1 and ψ_2. To return to the original variables ϑ_1 and ϑ_2, we are to make the transformation given by Equations 7.3.8.

Consider now another particular case, where the initial and terminal configurations are close to each other, i.e., $|\Delta\varphi_i| = |\varphi_i^1 - \varphi_i^0| \ll 1$, $i = 1, 2$. Linearizing Equations 7.2.3 in the neighborhood of the initial state (Equations 7.2.6) we obtain, under the conditions given by Equations 7.3.2,

$$(I_1 + 1)\ddot{\varphi}_1 + L\cos(\varphi_1^0 - \varphi_2^0)\ddot{\varphi}_2 = M_1 - M_2$$

$$I_2\ddot{\varphi}_2 + L\cos(\varphi_1^0 - \varphi_2^0)\ddot{\varphi}_1 = M_2 \qquad (7.3.17)$$

The change of variables

$$\psi_1 = [I_1 + 1 + L\cos(\varphi_1^0 - \varphi_2^0)]\varphi_1 + [I_2 + L\cos(\varphi_1^0 - \varphi_2^0)]\varphi_2$$

$$\psi_2 = L\cos(\varphi_1^0 - \varphi_2^0)]\varphi_1 + I_2\varphi_2 \qquad (7.3.18)$$

reduces Equations 7.3.17 to the form of Equations 7.3.9, whereas the boundary conditions (Equations 7.2.6 and 7.2.7) are reduced to the form of Equations 7.3.10 and 7.3.11.

Therefore, in terms of the variables in Equations 7.3.18, the solution of the optimal control problem in case $|\Delta\varphi_i| \ll 1$, $i = 1, 2$, is given by the same Equations 7.3.12, 7.3.13, 7.3.15, and 7.3.16, as in case L = 0. To return to the original variables, φ_1 and φ_2, we are to substitute Ψ_1 and Ψ_2 from Equations 7.3.18 into these relationships.

Note that if we apply the approach of Section 7.3.2 to our special cases L = 0 and $|\Delta\varphi_i| \ll 1$, we get the optimal controls obtained above in Section 7.3.3. Indeed, the total number of switches in controls given by Equations 7.3.3 are the same as for the optimal controls, and the instants of switches are in both cases found from the same Equations 7.3.5 and 7.3.6.

In the cases L = 0 and $|\Delta\varphi_i| \ll 1$, there exist two control laws of the class described by Equations 7.3.3 which lead to the same time of motion. Indeed, for $T_1 \geq T_2$, the control torque M_1 is determined uniquely by Equation 7.3.12, whereas for M_2 there are two possibilities, with $\beta = 0$ and $\beta = 1$ in Equation 7.3.13. Similarly, if $T_1 \leq T_2$, then M_2 is unique, while there are two possible controls M_1.

In our particular cases (L = 0 and $|\Delta\varphi_i| \ll 1$), we can choose any one of the two equivalent optimal control laws. In other words, we can limit ourselves to four types of controls from the eight variants presented in Equations 7.3.4, for instance, to k = 1, 2, 3, 4. As mentioned in Section 7.3.2, we acted in a similar way in the general case too. We can expect that, if L or $\Delta\varphi_i$ are small enough, our semi-inverse approach provides suboptimal controls close to the optimal ones.

7.3.4. NUMERICAL EXAMPLE

Consider a manipulator whose geometrical and inertial parameters (L_1, L, I_1, I_2, m_2) and upper bounds on control torques (M_{10} and M_{20}) are given by

$$L_1 = 1 \text{ m}, \qquad L = 0.5 \text{ m},$$

$$I_1 = I_2 = \frac{10}{3} \text{ kg·m}^2, \qquad m_2 = 10 \text{ kg}$$

$$M_{10} = 2 \text{ N·m}, \qquad M_{20} = 1 \text{ N·m} \qquad (7.3.19)$$

The orders of magnitudes of the parameters in Equations 7.3.19 correspond to typical industrial robots. The links of the robot are identical homogeneous bars. The boundary conditions are taken as follows:

$$\varphi_1(0) = 0, \qquad \dot{\varphi}_1(0) = 0, \qquad \varphi_2(0) = 0, \qquad \dot{\varphi}_2(0) = 0$$

$$\varphi_1(T) = \varphi_1^1 = -24°, \qquad \dot{\varphi}_1(T) = 0,$$

$$\varphi_2(T) = \varphi_2^1 = 72°, \qquad \dot{\varphi}_2(T) = 0 \qquad (7.3.20)$$

Using the dimensionless variables (Equations 7.3.1) and the radian measure for angles, we can rewrite Equations 7.3.19 and 7.3.20 as follows:

$$L_1 = 1, \qquad L = 0.5, \qquad I_1 = I_2 = \frac{1}{3}, \qquad m_2 = 1,$$

$$M_{10} = 2, \qquad M_{20} = 1;$$

$$\varphi_1^1 = -0.42 \text{ rad}, \qquad \varphi_2^1 = 1.26 \text{ rad} \qquad (7.3.21)$$

The diagram corresponding to the parameters given by Equations 7.3.21 is presented in Figure 7.3.3. Thick solid lines are the boundaries between the domains Ω_k, $k = 1, 2, 3, 4$ (see Section 7.3.2); thin solid curves show the lines of equal values of τ_1 and $\tau_* = \tau_2 - 2\tau_1$; dashed and dotted curves correspond to the lines $\tau_0 = $ const and $T = $ const, respectively.

Note that if $\varphi_1(0) = 0$ and $\varphi_2(0) = 0$, the diagrams for the control parameters are symmetric with respect to the origin of the plane (φ_1^1, φ_2^1), i.e., $\tau_i(-\varphi_1^1, -\varphi_2^1) = \tau_i(\varphi_1^1, \varphi_2^1)$, $i = 0, 1, 2$; $T(-\varphi_1^1, -\varphi_2^1) = T(\varphi_1^1, \varphi_2^1)$.

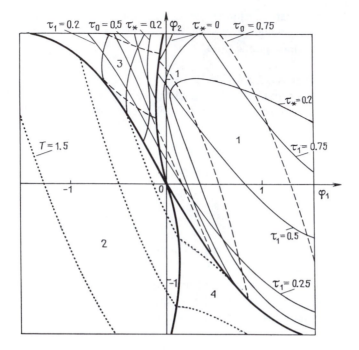

FIGURE 7.3.3.

This follows from the fact that Problem 7.1 with $\varphi_1(0) = 0$, $\varphi_2(0) = 0$, $\varphi_1(T) = \varphi_1^1$, $\varphi_2(T) = \varphi_2^1$ converts into the problem with $\varphi_1(T) = -\varphi_1^1$, $\varphi_2(T) = -\varphi_2^1$ if the variables φ_i and M_i, $i = 1, 2$, are changed for $-\varphi_i$ and $-M_i$. For this reason, we do not plot lines of equal levels for all parameters τ_0, τ_1, τ_*, and T in all domains Ω_k in Figure 7.3.3. In the domains Ω_1 and Ω_3, only the lines for τ_0, τ_1, and τ_* are shown, while the lines for T are plotted in the domains Ω_2 and Ω_4. The curves for τ_0 and T are drawn with the intervals $\Delta\tau_0 = 0.25$ and $\Delta T = 0.5$. The curves for τ_1 and τ_* are plotted with the intervals $\Delta\tau_1 = 0.21$ and $\Delta\tau_* = 0.1$ in the domain Ω_1, while in the domain Ω_2 the lines for τ_1 and τ_* equal to 0.2, 0.3, and 0.4 are given. All parameters change monotonically between the corresponding lines.

Using the diagram, we find that in our example the point with the coordinates $\varphi_1^1 = -0.42$ rad, $\varphi_2^1 = 1.26$ rad (see Equations 7.3.21) belongs to the domain Ω_3. Therefore, we should choose the control of the type $k = 3$ (see Equations 7.3.4). By means of interpolation between the curves in Figure 7.3.3, we determine the dimensionless parameters:

$$\tau_1 = 0.36, \qquad \tau_2 = 0.95, \qquad \tau_0 = 0.48, \qquad T = 1.18$$

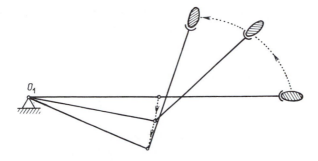

FIGURE 7.3.4.

Having returned to dimensional variables according to Equations 7.3.1 and 7.3.19, we obtain the dimensional values of the switching instants and the time of motion:

$$\tau_1 = 1.14 \text{ s}, \qquad \tau_2 = 3.00 \text{ s}, \qquad \tau_0 = 1.52_s, \qquad T = 3.72 \text{ s}$$

A diagrammatic representation of the corresponding motion of the manipulator is shown in Figure 7.3.4. Three consecutive configurations of the manipulator for $t = 0$, $t = 1.8$ s, and $t = T = 3.72$ s are given here. The dotted curves depict the trajectories of the gripper and the elbow. Of course, our solution is approximate; its accuracy depends on the density of lines in the diagram.

   To obtain more accurate solutions, we can replace graphic representations by numerical data stored in the computer memory. The semi-inverse approach described above provides suboptimal controls which can serve as an initial approximation for more accurate numerical methods.

7.4. OPTIMAL MOTIONS OF KINEMATICALLY REDUNDANT INDUSTRIAL ROBOTS

7.4.1. STATEMENT OF THE PROBLEM

   Let us consider a manipulator with n degrees of freedom that correspond to generalized coordinates $q_1, \ldots, q_n$ (angles of rotation and linear displacements of the links). The generalized coordinates uniquely determine the manipulator configuration as well as the position and orientation of a load in the gripper. The vector $q = (q_1, \ldots, q_n)$ is subject to the constraint $q \in Q$, where the domain Q specifies the set of kinematically admissible configurations in n-dimensional q-space. The position and orientation of the load is described by the vector $r = (r_1, \ldots, r_m)$, $m \leq 6$, including, for example, the coordinates of the mass center of the load and Euler's angles defining its orientation. Some of these coordinates may be not essential, and then $m < 6$. For example, we have $m = 3$ for planar motion of the load. If, in addition, the load is a body of revolution with its axis perpendicular to the plane of motion, we can assume $m = 2$. The generalized coordinates of the load are uniquely expressed through the vector q:

$$r = f(q), \qquad q \in Q, \qquad q = (q_1, \ldots, q_n)$$

$$r = (r_1, \ldots, r_m), \qquad f = (f_1, \ldots, f_m), \qquad m \leq n \qquad (7.4.1)$$

The specific form of the function f depends on the manipulator kinematics and the choice of the generalized coordinates q, r. In what follows, we consider the case of a kinematically redundant manipulator, and therefore, $m < n$.

Suppose the manipulator is driven so that each degree of freedom moves independently, and the time $T_i(q_i^0, q_i^1)$ during which the coordinate q_i changes from the value q_i^0 to q_i^1 depends only on $\Delta q_i = |q_i^0 - q_i^1|$ and is independent of other degrees of freedom and the mass of the load. Therefore, we have

$$T_i = T_i(\Delta q_i), \qquad \Delta q_i = |q_i^0 - q_i^1|, \qquad i = 1, \ldots, n \qquad (7.4.2)$$

Thus, the motion of the manipulator is completely defined by its initial and terminal configurations q^0 and q^1, provided that all degrees of freedom are engaged simultaneously. We also suppose that the drives can stop practically instantaneously. The time of motion is given by

$$T = \max_i T_i(\Delta q_i), \qquad i = 1, \ldots, n \qquad (7.4.3)$$

Consider two problems of minimizing the time of a robotic operation.

Problem 7.4. The initial configuration of the manipulator $q^0 \in Q$ and the desired terminal position r^1 of the load are given. Find the terminal configuration of the manipulator $q_*^1 \in Q$ that satisfies the condition $r^1 = f(q_*^1)$ and corresponds to the transfer of the load from the position $r^0 = f(q_0^1)$ to the position r^1 in the minimal time T_0:

$$T_0 = T(q^0, q_*^1) = \min_{q^1} T(q^0, q^1), \qquad f(q^1) = r^1, \qquad q^1 \in Q \quad (7.4.4)$$

Problem 7.5. The initial r^0 and terminal r^1 positions of the load are given. Find the initial and terminal configurations of the manipulator, $q_*^0 \in Q$ and $q_*^1 \in Q$, satisfying the conditions $r^i = f(q_*^i)$, $i = 0, 1$, and providing the transfer of the load from the position r^0 to the position r^1 in the minimal time T^0:

$$T^0 = T(q_*^0, q_*^1) = \min_{q^0, q^1} T(q^0, q^1)$$

$$f(q^j) = r^j, \qquad q^j \in Q, \qquad j = 0, 1 \qquad (7.4.5)$$

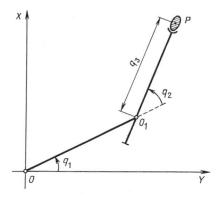

FIGURE 7.4.1.

To solve Problems 7.4 and 7.5 for concrete robots, we are to fulfill the following steps:

a.   Define the function f and the domain Q.
b.   Check the assumption of the independence of motions for different degrees of freedom and define the functions $T_i(\Delta q_i)$ (see Equation 7.4.2).
c.   Determine the minima (Equations 7.4.4 and 7.4.5).

Step a reduces to the analysis of the robot kinematics, step b requires experimental investigations, and step c consists in solving nonlinear programming problems. Below we solve Problems 7.4 and 7.5 for concrete industrial robots.

7.4.2. OPTIMAL TRANSPORT MOTIONS OF ROBOT UNIVERSAL-5

This robot consists of a fixed base, a platform, an arm mounted on the platform, and a gripper. The manipulator has six degrees of freedom which correspond to the vertical displacement of the platform, its rotation about the vertical axis, the horizontal translation of the arm, its rotation about the axis parallel to the rotation axis of the platform, and two rotations of the gripper, about the horizontal axis and about the axis perpendicular to the arm. The four first degrees of freedom perform transport motions and are driven by independent electric actuators. Orientation motions of the gripper are controlled by pneumatic drives.

We confine ourselves to the motions of the gripper in a horizontal plane with three active degrees of freedom: the extension of the arm and rotations of the platform and arm about vertical axes. Let these axes intersect the horizontal plane OXY at the points O and O_1, respectively (Figure 7.4.1). We take the angle q_1 between the line OO_1 and the axis OX, the angle q_2 between the line OO_1 and the axis of the arm O_1P, and the distance q_3 from the point

O_1 to the center of the gripper P, as the generalized coordinates of the manipulator.

The design of robot Universal-5 imposes the following constraints:

$$q \in Q = \{15° \leq q_1 \leq 345°; \qquad |q_2| \leq 120°;$$

$$0.43 \text{ m} \leq q_3 \leq 1.06 \text{ m}\} \qquad (7.4.6)$$

Let us describe the position of the load by the coordinates x and y of the gripper center P. We are not interested here in the orientation of the load. This is justified, for instance, in case the load is a body of revolution, its axis being kept vertical.

In our case kinematic Equations 7.4.1 are given by

$$x = e \sin q_1 + q_3 \sin(q_1 + q_2)$$

$$y = e \cos q_1 + q_3 \cos(q_1 + q_2)$$

$$r = (x, y), \qquad e = OO_1 = 0.5 \text{ m} \qquad (7.4.7)$$

Here, e is the distance between rotation axes of the platform and the arm.

To investigate the dynamics of the manipulator drives and to find out the functions $T_i(\Delta q_i)$ (Equation 7.4.2) we performed a series of experiments. For each of the three coordinates of the robot, q_1, q_2, and q_3, we measured $T_i(\Delta q_i)$, $i = 1, 2, 3$, provided the rest degrees of freedom were not active. The manipulator was driven for the desired displacement Δq_i and returned to the initial state. To enhance the measurement accuracy, we repeated this motion ten times. The experiments show that for the examined degrees of freedom the time of motion depends on $\Delta q_i = |q_i^0 - q_i^1|$ rather than on q_i^0 and q_i^1 separately. The functions $T_i(\Delta q_i)$ increase monotonically. At $\Delta q_i = 0$, they jump from $T_i = 0$ to $T_i = 0.5$ s, due to the sluggishness of the mechanical part of the control unit.

The graphs of the functions $T_i(\Delta q_i)$, $i = 1, 2, 3$ are given in Figure 7.4.2. The curve 1 corresponds to $T_1(\Delta q_1)$ for $q_2 = 0$, $q_3 = 1.06$ m. The influence of the load mass m_0 on the time of the platform rotation turned out to be insignificant. For instance, when the gripper is unloaded ($m_0 = 0$), the time of rotation of the platform by the angle $\Delta q_1 = 180°$ is $T_1 = 4.8$ s, while for $m_0 = 6$ kg (it is 1 kg greater than the nominal load-carrying capacity for this robot) we have $T_1 = 4.9$ s. The change of the inertia moment of the arm due to the change of the coordinate q_2 causes not more than 3% change of the time $T_1(\Delta q_1)$.

The solid curve 2 shows the function $T_2(\Delta q_2)$ for the unloaded manipulator with $q_3 = 1.06$ m. The dashed line corresponds to $m_0 = 6$ kg. With this load, the time of the motion increases not more than by 7%. The dashed-and-dotted curve gives $T_2(\Delta q_2)$ for $q_3 = 0.79$ m. The change of the arm

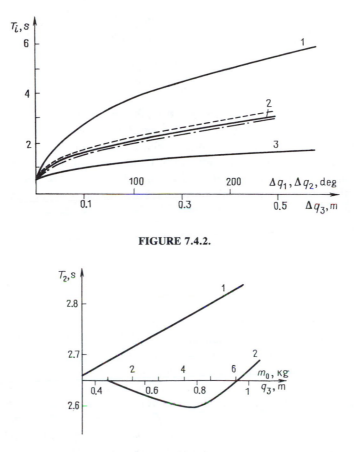

FIGURE 7.4.2.

FIGURE 7.4.3.

length causes the change of the inertia moment of the arm, but the rotation time T_2 changes not more than by 4%.

The curve 3 in Figure 7.4.2 presents $T_3(\Delta q_3)$. The load of the mass 6 kg increases the time of motion for this degree of freedom not more than by 2%.

Figure 7.4.3 shows the dependence of the time T_2 on the mass m_0 (the curve 1) and on the coordinate q_3, for the unloaded gripper (the curve 2). The first dependence is linear, while the second reveals the minimum $T_2 \approx 2.6$ s, at $q_3 \approx 0.79$ m, because the moment of inertia of the arm about the axis O_1 is minimal for $q_3 \approx 0.79$ m.

The experimental results confirm the assumptions about the properties of functions $T_i(\Delta q_i)$ in Equation 7.4.2. Therefore, we can solve Problems 7.4 and 7.5 following the general scheme of Section 7.4.1.

In our case, $n = 3$ and $m = 2$. To determine the generalized coordinates uniquely for given x and y, we are to specify one of q_i, e.g., q_1. Then q_2 and q_3 are expressed from Equation 7.4.7 as follows:

$$q_2 = q_2(q_1) = \tan^{-1}\left(\frac{x - e \sin q_1}{y - e \cos q_1}\right) - q_1 + \pi N$$

$$q_3 = q_3(q_1, q_2) = \frac{x - e \sin q_1}{\sin(q_1 + q_2)} \tag{7.4.8}$$

Here, the integer N is to be determined from the condition $q \in Q$ (see Equation 7.4.6). It follows from Equations 7.4.8 and 7.4.6 that N is found uniquely and can assume one of the following values: $N = -1, 0, 1, 2, 3$. If the condition $q \in Q$ is not fulfilled for any of these values of N, then there is no admissible configuration for the chosen q_1. Problem 7.4 for our robot is reduced to the constrained minimization of a function of a single variable, q_1^1:

$$T_0 = T(q^0, q_*^1) = \min_{q_1^1} T(q^0, q^1)$$

$$q^1 = \{q_1^1, q_2(q_1^1), q_3^1(q_1^1, q_2(q_1^1))\} \in Q \tag{7.4.9}$$

Problem 7.4 was solved numerically. The functions $T_i(\Delta q_i)$ shown in Figure 7.4.2 were introduced into a computer: 67 points of the curve 1 for $T_1(\Delta q_1)$, with the step $5°$ in Δq_1; 49 points of the curve 2 for $T_2(\Delta q_2)$, with the step $5°$ in Δq_1; and 64 points of the curve 3 for $T_3(\Delta q_3)$, with the step 0.01 m in Δq_3. Between these nodal points, the functions $T_i(\Delta q_i)$ were determined by linear interpolation with the accuracy comparable with that of the experiments. The function $T(q^0, q^1)$ was calculated according to Equation 7.4.3, with $n = 3$. The minimum of $T(q^0, q^1)$ in Equation 7.4.9 was determined by a direct search for the variable q_1^1, with the step $1°$. For each fixed q_1^1, the values q_2^1 and q_3^1 were calculated according to Equations 7.4.8.

Example 7.1. Let the initial configuration q^0 and the desired position $r^1 = (x^1, y^1)$ of the gripper P be specified as follows:

$$q^0: q_1^0 = 90°, \qquad q_2^0 = 0, \qquad q_3^0 = 0.83 \text{ m}$$
$$r^1: x^1 = 0, \qquad y^1 = -1.4 \text{ m}$$

It is required to transfer the gripper from the initial position $r^0: x^0 = 1.4$ m, $y^0 = 0$, corresponding to the configuration q^0, to the terminal position. This transfer can be carried out by rotating the platform clockwise by $90°$. It takes the time $T \approx 3.6$ s. This motion is not optimal. The optimal motion takes the time $T_0 \approx 2.8$ s and corresponds to the terminal configuration of the manipulator $q_*^1 = (137°, 65°, 1.06$ m$)$. The optimal motion results in 22% reduction of the operational time.

Problem 7.5 for robot Universal-5 is reduced to the minimization of a function of two variables, q_1^0 and q_1^1:

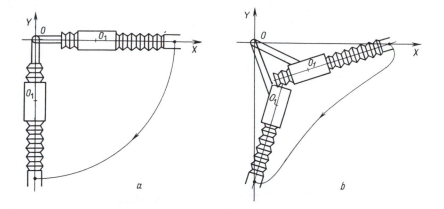

FIGURE 7.4.4.

$$T^0 = T(q^0_*, q^1_*) = \min_{q^0_1, q^1_1} T(q^0, q^1)$$

$$q^j = \{q^j_1, q_2(q^j_1), q_3(q^j_1, q_2(q^j_1))\} \in Q, \qquad j = 0, 1 \qquad (7.4.10)$$

The minimum in Equation 7.4.10 was obtained by a direct search for the two variables q^0_1 and q^1_1 within the limits given by Equation 7.4.6, with the step $1°$.

Example 7.2. The initial and terminal positions of the gripper P are given by

$$r^0: x^0 = 1.4 \text{ m}, \qquad y^0 = 0; \qquad r^1: x^1 = 0, \qquad y^1 = -1.4 \text{ m}$$

These positions are the same as in Example 7.1. However, here, unlike Example 7.1., the initial configuration q^0 of the manipulator is not specified beforehand.

The solution of Problem 7.4 results in the following optimal initial and terminal configurations:

$$q^0_* = (119°, -46°, 0.94 \text{ m}), \qquad q^1_* = (137°, 65°, 1.06 \text{ m})$$

Figure 7.4.4 shows the nonoptimal (a) and optimal (b) motions for the given example. The nonoptimal motion is the rotation of the platform by the angle $90°$ about the axis O_1; it takes $T = 3.6$ s (see Example 1). The optimal motion takes $T^0 = 2.1$ s, that is 25% less than the optimal time $T_0 = 2.8$ s from Example 7.1, where the initial configuration is fixed, and 41% less as compared with the nonoptimal motion.

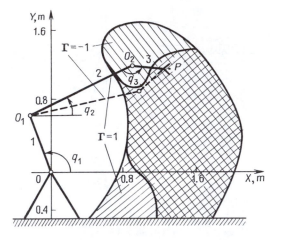

FIGURE 7.4.5.

7.4.3. OPTIMAL MOTIONS OF ROBOT RPM-25

Anthropomorphic manipulation robot RPM-25 has five degrees of freedom and consists of a base, three links, and a gripper (Figure 7.4.5). We will call the links 1, 2, and 3 an upper arm, a forearm, and a hand, respectively. The manipulator can rotate about the vertical axis, the links 1, 2, and 3 can rotate about parallel horizontal axes, and the gripper can rotate about the axis of the hand. All the degrees of freedom are controlled independently by means of electric drives.

Below, we consider only three degrees of freedom of the robot corresponding to swings (rotations about horizontal axes) of the upper arm, forearm, and hand. Using these degrees of freedom, we can transfer the gripper to any point in a vertical plane inside the operation volume of the manipulator.

Let us introduce the coordinate frame OXY whose origin lies on the swing axis of the upper arm, the axis OX is horizontal, and the axis OY is pointed vertically (see Figure 7.4.5). For the generalized coordinates of the manipulator, we take the angles q_1 and q_2 formed by the upper arm and the forearm, respectively, with the OX axis, and the angle q_3 between the forearm and the hand (see Figure 7.4.5). Robot RPM-25 is designed so that if only the drive of the upper arm is engaged, the angle q_1 changes while q_2 is kept constant, i.e., the forearm moves translationally.

The domain Q of possible configurations of manipulator RPM-25 is given by

$$Q = \{[45° \leq q_1 \leq 90°, \quad -50° \leq q_2 \leq -25° + \frac{5}{9} q_1]$$

$$\cup \quad [90° \leq q_1 \leq 135°, \quad -150° + \frac{10}{9} q_1 \leq q_2 \leq 25°]\} \quad (7.4.11)$$

$$\cap \quad [90° \leq q_3 \leq 270°]$$

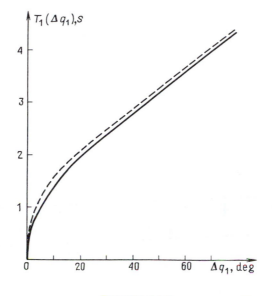

FIGURE 7.4.6.

The corresponding working zone in the plane OXY is shaded in Figure 7.4.5. We will characterize the position of the load by the Cartesian coordinates, x and y, of the center of the gripper P. The Cartesian coordinates x, y and the generalized coordinates q_1, q_2, q_3 are related by

$$x = l_1\cos q_1 + l_2\cos q_2 - l_3\cos(q_2 + q_3)$$

$$y = l_1\sin q_1 + l_2\sin q_2 - l_3\sin(q_2 + q_3)$$

$$l_1 = |OO_1| = 0.7 \text{ m}, \qquad l_2 = |O_1O_2| = 1.3 \text{ m}, \qquad (7.4.12)$$

$$l_3 = |O_2P| = 0.4 \text{ m}$$

where l_1, l_2, and l_3 are the lengths of the manipulator links. Thus, as in Section 7.4.2, we have $n = 3$, $m = 2$, i.e., the manipulator is kinematically redundant.

We found experimentally the functions $T_i(\Delta q_i)$ for all generalized coordinates, q_1, q_2, and q_3. We investigated also, how these functions were influenced by the interaction of different degrees of freedom and by the mass of the payload. These experiments are similar to those described in Section 7.4.2.

In Figures 7.4.6 through 7.4.8, solid curves present the functions $T_i(\Delta q_i)$, $i = 1, 2, 3$ for the unloaded gripper. Dashed lines in Figures 7.4.6 and 7.4.7 correspond to $T_i(\Delta q)$, $i = 1,2$ for the manipulator carrying a load of the maximal admissible mass $m_0 = 25$ kg. The influence of the load on $T_3(\Delta q_3)$ is insignificant. The experiments indicate that the functions $T_i(\Delta q_i)$ for each

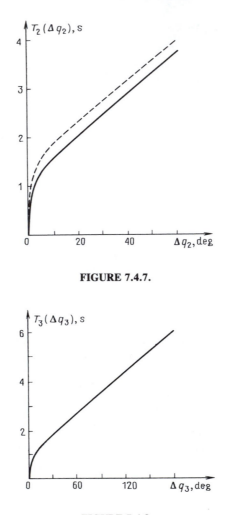

FIGURE 7.4.7.

FIGURE 7.4.8.

degree of freedom can be regarded, with high accuracy, as independent of other degrees of freedom, as well as of the load mass. Therefore, for our manipulator, the approach of Section 7.4.1 is applicable.

By virtue of the kinematic redundancy the coordinates x and y do not determine uniquely the vector q. Let us fix x and y, and, in addition, the generalized coordinate q_1. Then, there exist not more than two values of q_2 (q_2^+ and q_2^-) such that each of the pairs (q_1, q_2^+) and (q_1, q_2^-) uniquely determines the third coordinate q_3 (q_3^+ and q_3^-, respectively), both triads (q_1, q_2^+, q_3^+) and (q_1, q_2^-, q_3^-) corresponding to the same position of the gripper (x, y). This is explained by the fact that for a given q_1, two relative positions of the forearm and hand can exist for a fixed location of the gripper P. As in Section 7.2, we characterize the two possible configurations by the parameter $\Gamma = \pm 1$; $\Gamma = 1$ ($90° \le q_3 \le 180°$) corresponds to the angles q_2^+ and q_3^+, while $\Gamma = -1$ ($180° \le q_3 \le 270°$) corresponds to q_2^- and q_3^-. In Figure 7.4.5, the solid and dashed lines show configurations with $\Gamma = 1$ and $\Gamma = -1$, respectively. The coordinates q_2^+, q_3^+ and q_2^-, q_3^- can be found from Equations 7.4.12 and expressed as follows:

$$q_2^{\pm} = \tan^{-1}\frac{b}{a} + \Gamma \cos^{-1}\frac{c^2 + l_2^2 - l_3^2}{2l_2 c}$$

$$q_3^{\pm} = q_3(x, y, q_1, \Gamma) = \begin{cases} \cos^{-1}(l_2^2 + l_3^2 - c^2)/(2l_2 l_3), & \Gamma = 1 \\ 2\pi - q_3(x, y, q_1, 1), & \Gamma = -1 \end{cases}$$

$$a = x - l_1 \cos q_1, \qquad b = y - l_1 \sin q_1, \qquad c^2 = a^2 + b^2 \qquad (7.4.13)$$

Suppose the coordinates x and y are known. The configuration q is determined uniquely, if we specify q_1 and Γ so that the vector (q_1, q_2, q_3) calculated according to Equations 7.4.13 belongs to the domain Q from Equation 7.4.11.

Numerical analysis shows that if the gripper P is located in the region covered by the double hatching in Figure 7.4.5, then for any $\Gamma = \pm 1$, one can find the angle q_1 so that $q \in Q$. However, if the gripper lies in the region covered by the single hatching, the admissible angle q_1 can be found only for $\Gamma = 1$ or $\Gamma = -1$.

Problem 7.4 for robot RPM-25 reduces the two constrained minimization problems for a function of one variable,

$$T_0 = T(q^0, q_*^1) = \min_{q_1^1} T(q^0, q^1)$$

$$q^1 \in Q, \qquad \Gamma^1 = \pm 1 \qquad (7.4.14)$$

while Problem 7.5 reduces to four minimization problems for a function of two variables:

$$T^0 = T(q_*^0, q_*^1) = \min_{q_1^0, q_1^1} T(q^0, q^1)$$

$$q^j \in Q, \qquad \Gamma^j = \pm 1, \qquad j = 0, 1 \qquad (7.4.15)$$

Here, Γ^0 and Γ^1 are the types of the manipulator configurations at the initial and terminal positions; q_2 and q_3 are expressed through q_1 and Γ by Equations 7.4.13.

To determine T_0, one should solve Equations 7.4.14 for $\Gamma = 1$ and $\Gamma = -1$ and then select the solution corresponding to the lesser time. To determine T^0, one should find the minima in Equation 7.4.15 for four possible combinations of $\Gamma^0 = \pm 1$ and $\Gamma^1 = \pm 1$, and then take the least time.

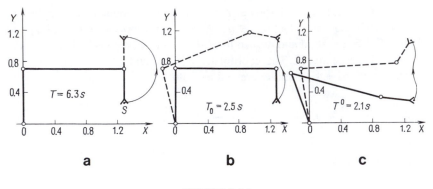

FIGURE 7.4.9.

The minima (Equations 7.4.14 and 7.4.15) were obtained by means of direct search, analogously to Section 7.4.2. The functions $T_i(\Delta q_i)$ (see Figures 7.4.6 to 7.4.8) were fed into a computer as arrays of the experimental data. Between the nodal points, the values of T_i were determined by linear interpolation. The time $T(q^0, q^1)$ was calculated according to Equation 7.4.3. Below, we present some examples of calculations of optimal motions.

Example 7.3. Let the initial configuration of the manipulator and the terminal position of the load be given by

$$q^0: q_1^0 = 90°, \qquad q_2^0 = 0, \qquad q_3^0 = 90°$$

$$r^1: x^1 = 1.3 \text{ m}, \qquad y^1 = 1.1 \text{ m} \qquad\qquad (7.4.16)$$

It is required to transfer the gripper P from the initial position $r^0: x^0 = 1.3$ m, $y^0 = 0.3$ m, corresponding to q^0 in Equations 7.4.16, to the terminal position r^1. This transfer can be performed by rotating the hand by the angle 180° (see Figure 7.4.9a). It takes the time $T = 6.3$ s. The optimal motion obtained as the solution of Problem 7.4 (see Equation 7.4.14) takes the time $T_0 = 2.5$ s which is 60% less than T. The optimal terminal configuration of the manipulator is $q_*^1 = (113°, 25°, 143°)$; see Figure 7.4.9b.

Example 7.4. The initial and terminal positions of the load are specified as follows:

$$r^0: x^0 = 1.3 \text{ m}, \qquad y^0 = 0.3 \text{ m};$$

$$r^1: x^1 = 1.3 \text{ m}, \qquad y^1 = 1.1 \text{ m} \qquad\qquad (7.4.17)$$

The solution of Problem 7.5 (see Equation 7.4.15) provides the optimal initial and terminal configurations of the manipulator given by

$$q_*^0 = (122°, -12°, 189°), \qquad q_*^1 = (109°, 5°, 229°)$$

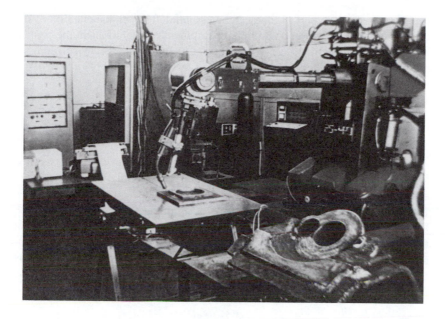

FIGURE 7.4.10.

The optimal motion, shown in Figure 7.4.9c, takes the time $T^0 = 2.1$ s, which is 16% less than T_0 and 67% less than T (see Example 7.3).

The results presented in Section 7.4.3 were used for planning optimal motions of robot RPM-25 incorporated in the robotic system for welding.[102] The general view of this system is given in Figure 7.4.10.

7.4.4. CONCLUDING REMARKS

The approach described above makes it possible to plan time-optimal motions for kinematically redundant manipulation robots. This approach is applicable if the motions corresponding to different degrees of freedom do not practically affect each other. This assumption is satisfied with a sufficient accuracy for many industrial robots equipped with powerful drives and reduction gears with high gear ratios. The calculation of the optimal initial and terminal configurations is comparatively simple and does not require mathematical models in the form of differential equations. Only certain experimental measurements should be carried out. The optimal motions obtained as a result of our approach can provide a considerable gain in operational time. This gain is achieved due to the rational combination of motions for different degrees of freedom.

The optimal motions calculated for industrial robots Universal-5 and RPM-25 were implemented experimentally. The operational time appeared to be close to the predicted optimum.

The optimization is most expedient if the robot repeats a certain operation in some cyclic technological process many times. In the case of the "bottleneck'' operation, savings in time can be essential. The implementation of the optimal motions requires no additional equipment or alterations in the control hardware. These motions can be performed using standard control units of robots.

7.5. OPTIMAL CONTROL OF ELECTROMECHANICAL ROBOTS

7.5.1. EQUATIONS OF MOTION

Consider a manipulator that has n degrees of freedom and consists of n links connected by revolute or prismatic joints. Each link of the robot is an absolutely rigid body. The position of the kth link with respect to the $(k - 1)$st is characterized by the relative angle of rotation (in the case of a revolute joint) or by a relative displacement (in the case of a prismatic joint). We take these angles and displacements as generalized coordinates $q_1, \ldots, q_n$, determining the configuration of the robot. The motion of the manipulator is governed by the Lagrangian Equations 2.3.27, which can be rewritten as follows:

$$a(q)\ddot{q} + S(q, \dot{q}) = Q \qquad (7.5.1)$$

Here, q is the vector of generalized coordinates, $A(q)$ is the symmetric positive-definite n $\times$ n matrix of the kinetic energy, $Q = (Q_1, \ldots, Q_n)$ is the vector of generalized control forces created by drives. The role of these forces is played by the torques about the axes of revolute joints and by the forces in the directions of displacements in the case of prismatic joints. The vector $S(q, \dot{q})$ in Equation 7.5.1 is given by

$$S_i(q, \dot{q}) = \frac{1}{2} \sum_{j,k=1}^{n} \left(\frac{\partial a_{ij}}{\partial q_k} + \frac{\partial a_{ik}}{\partial q_j} - \frac{\partial a_{jk}}{\partial q_i} \right) \dot{q}_j \dot{q}_k - Q'_i,$$

$$i = 1, \ldots, n \qquad (7.5.2)$$

Here, a_{ij} are elements of the matrix A, and Q'_i are active generalized forces other than the control ones, for example, gravity, resistance, friction, etc.

Suppose that each control torque or force Q_i is produced by a separate DC electric motor. The motors of industrial robots usually have gear trains with large gear ratios $N_i \gg 1$. The angular velocity ω_i of the rotor of the ith motor and the respective generalized velocity are related by

$$\omega_i = N_i \dot{q}_i, \qquad N_i \gg 1 \qquad (7.5.3)$$

Let us compile the equation for rotation of the rotor of the ith motor, neglecting the influence of the motion of the robot on this rotation (this is admissible for $N_i \gg 1$). We obtain

$$J_i \dot{\omega}_i = -b_i \omega_i + M_i - N_i^{-1} Q_i, \qquad i = 1, \dots, n \qquad (7.5.4)$$

Here, J_i is the moment of inertia of the rotating parts of the ith motor about its axis, the term $b_i \omega_i$ is the torque due to the mechanical resistance, b_i is a positive constant coefficient, and M_i is the electromagnetic torque created by the motor. The torque M_i is proportional to the electric current j_i of the ith motor:

$$M_i = c_i j_i, \qquad c_i > 0 \qquad (7.5.5)$$

Here, c_i is a constant. The equation of balance of voltages in the circuit of the ith motor has the form

$$L_i dj_i/dt + R_i j_i + h_i \omega_i = u_i \qquad (7.5.6)$$

Here, L_i is the inductance, R_i is the electric resistance, h_i is a constant coefficient, and u_i is the electric voltage applied to the ith motor. The term $L_i dj_i/dt$ in Equation 7.5.6 is usually small compared with the other terms and can be omitted (see also Section 3.1.4). Then, from Equations 7.5.5 and 7.5.6 we obtain

$$M_i = c_i R_i^{-1}(u_i - h_i \omega_i) \qquad (7.5.7)$$

Substituting ω_i from Equation 7.5.3 and M_i from Equation 7.5.7 into Equation 7.5.4, we obtain

$$J_i N_i \ddot{q}_i + b_i N_i \dot{q}_i = c_i R_i^{-1}(u_i - h_i N_i \dot{q}_i) - N_i^{-1} Q_i$$

From this equation we determine Q_i:

$$Q_i = -N_i^2 J_i \ddot{q}_i - N_i^2 (b_i + c_i h_i R_i^{-1}) \dot{q}_i + N_i c_i R_i^{-1} u_i, \qquad i = 1, \dots, n$$

We substitute the expression obtained for Q_i into the Lagrangian Equations 7.5.1. As a result, we obtain the equations of motion in the form

$$N_i^2 J_i \ddot{q}_i + (A\ddot{q})_i + N_i^2 (b_i + c_i h_i R_i^{-1}) \dot{q}_i + S_i(q, \dot{q})$$

$$= N_i c_i R_i^{-1} u_i, \qquad i = 1, \dots, n \qquad (7.5.8)$$

In this equation, we make the change of variables,

$$y_i = N_i q_i, \qquad i = 1, \ldots, n \qquad (7.5.9)$$

and divide the ith Equation 7.5.8 by N_i. We obtain

$$J\ddot{y} + GA(Gy)G\ddot{y} + \Lambda\dot{y} + GS = Ku$$

$$J = \text{diag}(J_i), \qquad G = \text{diag}(N_i^{-1}),$$

$$\Lambda = \text{diag}(\Lambda_i), \qquad \Lambda_i + b_i + c_i h_i R_i^{-1}$$

$$K = \text{diag}(K_i), \qquad K_i = c_i R_i^{-1}, \qquad u = (u_1, \ldots, u_n) \qquad (7.5.10)$$

Here, $\text{diag}(a_i)$ denotes the diagonal $n \times n$ matrix with diagonal elements equal to a_i. The Equation 7.5.10 can be represented in the form

$$A^*(y)\ddot{y} + S^* = Q^*, \qquad A^*(y) = J + GA(Gy)G$$

$$S^* = \Lambda\dot{y} + GS, \qquad Q^* = Ku \qquad (7.5.11)$$

Here, A^* is a symmetric positive-definite matrix. Thus, the equations of motion for our electromechanical robot are reduced to the form of the Lagrangian Equations 7.5.1.

Equations 7.5.10 and 7.5.11 contain the diagonal matrix G with small diagonal elements $N_i^{-1} \ll 1$. Hence, we can simplify these equations, neglecting the terms depending on G in Equation 7.5.10. Thus we obtain the following decoupled linear system:

$$J_i\ddot{y}_i + \Lambda_i\dot{y}_i = K_i u_i, \qquad i = 1, \ldots, n \qquad (7.5.12)$$

Returning to our original coordinates q_i (see Equations 7.5.9), we can rewrite Equations 7.5.12 as follows:

$$J_i N_i\ddot{q}_i + \Lambda_i N_i\dot{q}_i = K_i u_i, \qquad i = 1, \ldots, n \qquad (7.5.13)$$

This simplified system can be used as an approximate mathematical model of the electromechanical multilink manipulator.

The constraints

$$|u_i| \le u_i^0, \qquad i = 1, \ldots, n \qquad (7.5.14)$$

where u_i^0 are given constants, are usually imposed on the control voltages. Sometimes, the currents through the armature circuits of the motors should also be restricted,

$$|j_i| \leq j_i^0, \qquad i = 1, \ldots, n \qquad (7.5.15)$$

where j_i^0 are given bounds. Using Equations 7.5.5, 7.5.7, and 7.5.3, we can reduce the Equations 7.5.15 to the form

$$|u_i - h_i N_i \dot{q}_i| \leq R_i j_i^0, \qquad i = 1, \ldots, n \qquad (7.5.16)$$

Let us state the time-optimal control problem for our simplified model.

Problem 7.6. Find the feedback control $u(q, \dot{q})$ subject to constraints given by Equations 7.5.14 and 7.5.16 and driving the system governed by Equation 7.5.13 from an arbitrary initial state,

$$q(0) = q^0, \qquad \dot{q}(0) = \dot{q}^0 \qquad (7.5.17)$$

to a given terminal state,

$$q(T) = q^1, \qquad \dot{q}(0) = 0 \qquad (7.5.18)$$

in the minimal possible time T.

Since the Equations 7.5.13 as well as Equations 7.5.14 and 7.5.16 are decoupled, we can solve Problem 7.6 for each degree of freedom separately.

7.5.2. OPTIMAL CONTROL FOR A SYSTEM WITH ONE DEGREE OF FREEDOM

Consider Equations 7.5.13, 7.5.14, and 7.5.16 for the ith degree of freedom. Let us omit the subscript i and introduce the dimensionless variables and constants:

$$t' = hKJ^{-1}t, \qquad q' = h^2 KNJ^{-1}(u^0)^{-1}q$$

$$u' = (u^0)^{-1}u, \qquad \alpha = \Lambda K^{-1}h^{-1}, \qquad \eta = Rj^0(u^0)^{-1} \qquad (7.5.19)$$

Substituting Equations 7.5.19 into Equations 7.5.13, 7.5.14, and 7.5.16, we obtain

$$\ddot{q} + \alpha\dot{q} = u, \qquad |u| \leq 1, \qquad |u - \dot{q}| \leq \eta \qquad (7.5.20)$$

Here and henceforth the primes are omitted.

Let us confine ourselves to the case where the mechanical resistance in the motor is negligible. The resistance can be reduced significantly by using high-quality bearings and lubricants. If the resistance coefficient b can be neglected, we obtain from Equations 7.5.19 and 7.5.10 that $\alpha = 1$. Thus, our equations and constraints (7.5.20) can be rewritten as

$$\ddot{q} + \dot{q} = u \qquad (7.5.21)$$

$$|u| \leq 1, \qquad |u - \dot{q}| \leq \eta \qquad\qquad (7.5.22)$$

Equations 7.5.22 can be represented in the following equivalent form:

$$u^1(\dot{q}) \leq u \leq u^2(\dot{q})$$

$$u^1(\dot{q}) = \max(-1, \dot{q} - \eta), \qquad u^2(\dot{q}) = \min(1, \dot{q} + \eta) \quad (7.5.23)$$

The set bounded by Equations 7.5.23 is nonempty whenever

$$|\dot{q}| \leq 1 + \eta \qquad\qquad (7.5.24)$$

Let us introduce a new control variable v as follows:

$$u = f(\dot{q})v + g(\dot{q})$$

$$f(\dot{q}) = [u^2(\dot{q}) - u^1(\dot{q})]/2$$

$$g(\dot{q}) = [u^1(\dot{q}) + u^2(\dot{q})]/2 \qquad\qquad (7.5.25)$$

It stems from Equations 7.5.23 and 7.5.25 that the constraint given by Equations 7.5.23 is equivalent to $|v| \leq 1$. The functions $f(\dot{q})$ and $g(\dot{q})$ are continuous and piecewise continuously differentiable, their derivatives having jumps at the points $\dot{q} = \pm(1 - \eta)$. The function $f(\dot{q})$ is nonnegative in the domain specified by Equation 7.5.24, and $f(\dot{q}) > 0$ if Inequality 7.5.24 is strict.

By reducing Equation 7.5.21 to the set of first-order differential equations, we represent Problem 7.6 for our system with one degree of freedom as follows:

$$\dot{x}_1 = x_2, \qquad \dot{x}_2 = -x_2 + f(x_2)v + g(x_2) \qquad (x_1 = q) \quad (7.5.26)$$

$$x_1(0) = x_1^0 = q^0, \qquad x_2(0) = x_2^0 = \dot{q}^0 \qquad\qquad (7.5.27)$$

$$x_1(T) = x_1^1 = q^1, \qquad x_2(T) = 0 \qquad\qquad (7.5.28)$$

$$|v| \leq 1 \qquad\qquad (7.5.29)$$

$$|x_2| \leq 1 + \eta \qquad\qquad (7.5.30)$$

The time-optimal control problem (Equations 7.5.26 through 7.5.30) includes the state constraint (Equation 7.5.30). We will see that all optimal trajectories starting inside the domain given by Equation 7.5.30 never reach the boundaries of this domain. Hence, we can ignore the state constraint and use Pontryagin's maximum principle.

The Hamiltonian for our problem is

$$H = p_1 x_2 - p_2 x_2 + p_2 f(x_2) v + p_2 g(x_2) \qquad (7.5.31)$$

where p_1 and p_2 are adjoint variables. We have $f(x_2) > 0$ inside the domain given by Equation 7.5.30, hence the maximum of H with respect to v under the constraint specified by Equation 7.5.29 is attained at $v = \text{sign } p_2$. The adjoint system corresponding to the Hamiltonian (Equation 7.5.31) is given by

$$\dot{p}_1 = 0, \qquad \dot{p}_2 = -p_1 + [1 - f'(x_2) v - g'(x_2)] p_2 \qquad (7.5.32)$$

As mentioned above, the functions f' and g' have jumps on the lines $x_2 = \pm(1 - \eta)$ in the phase (x_1, x_2) plane. In this case, the maximum principle must be supplemented with additional conditions for the jumps of the adjoint variables on the lines of discontinuities. It can be shown that in our problem the optimal trajectories can only intersect the discontinuity lines $x_2 = \pm(1 - \eta)$ and do not contain segments lying on them. In this case, the additional conditions mentioned above imply that the adjoint variables are continuous. Hence, we can use the maximum principle in its conventional form as if the right-hand sides of Equations 7.5.26 are smooth.

Let us prove that the optimal control $v = \text{sign } p_2$ has not more than one switch.

First, we remark that p_2 cannot be identically zero on any time interval. Indeed, assuming $p_2 \equiv 0$ on some time interval we obtain from Equations 7.5.32 that $p_1 = \text{const} = 0$. Hence, the both adjoint variables are identically zero, in contradiction with the necessary optimality conditions.

Suppose now there are at least two instants of switch. Let t_1 and t_2 be the neighboring zeros of $p_2(t)$. Then it follows from Equations 7.5.32 that

$$\dot{p}_2(t_1) = \dot{p}_2(t_2) = -p_1 \qquad (7.5.33)$$

The derivatives $\dot{p}_2(t_1)$ and $\dot{p}_2(t_2)$ cannot have the same sign; therefore, Equations 7.5.33 can hold only if $p_1 = 0$. In this case Equations 7.5.32 imply that $p_2(t) \equiv 0$, and we again arrive at the contradiction with the maximum principle.

Now, let us construct the original trajectory leading from the initial state (Equations 7.5.27) to the terminal state (Equations 7.5.28). Integrating Equa-

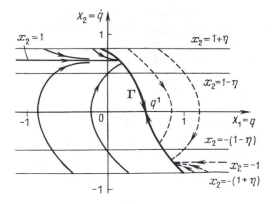

FIGURE 7.5.1.

tions 7.5.26 with constant $v = \pm 1$, we obtain the following expressions for
the phase trajectories:

$$x_1 = A_1 + v x_2^2/(2\eta) \quad \text{if} \quad -(1 + v\eta) \leq x_2 \leq 1 - v\eta$$

$$x_1 = A_2 - x_2 - v \ln(1 - v x_2) \quad \text{if} \quad v - \eta \leq x_2 \leq v + \eta \quad (7.5.34)$$

Two phase trajectories of those specified by Equations 7.5.34, corresponding to $v = 1$ and $v = -1$, pass through each point of the domain given
by Equation 7.5.30. They are shown by solid (for $v = 1$) and dashed (for $v
= -1$) lines in Figure 7.5.1. It is seen from this figure that for any initial
state (Equation 7.5.27) there is a unique trajectory with only one switch of
the control v that leads from any initial state (Equation 7.5.27) to the terminal
state (Equation 7.5.28). The switch of the bang-bang control can occur on
two half-trajectories (Equation 7.5.34), leading to the terminal state and corresponding to $v = \pm 1$. These half-trajectories form the switching curve Γ
shown in Figure 7.5.1 by a thick line. The feedback optimal control $v(x_1\, x_2)$
can be described as follows. The control is $v = -1$ above and to the right
of the curve Γ, and $v = 1$ below and to the left of Γ. On the curve Γ, we
have $v = -\text{sign } x_2$. We can see from Figure 7.5.1 that all obtained optimal
trajectories intersect the lines $x_2 = \pm(1 - \eta)$ where the right-hand sides of
Equations 7.5.26 are discontinuous. Moreover, all optimal trajectories do not
leave the domain $|x_2| \leq 1 + \eta$. Thus, all assumptions made above are justified,
and our solution is completed.

Returning to the variables q, $\dot{q}$, and u according to Equations 7.5.25 and
7.5.26, we can express the obtained feedback optimal control as follows:

$$u(q, \dot{q}) = u^1(\dot{q}), \quad \text{if} \quad \psi < 0 \quad \text{or} \quad \psi = 0, \quad \dot{q} > 0$$

$$u(q, \dot{q}) = u^2(\dot{q}), \quad \text{if} \quad \psi > 0 \quad \text{or} \quad \psi = 0, \quad \dot{q} < 0 \quad (7.5.35)$$

Here, $\psi(q, \dot{q})$ is the switching function given by

$$\psi(q, \dot{q}) = q^1 - q - \dot{q}^2 \operatorname{sign} \dot{q}/(2\eta), \quad \text{if} \quad 0 < \eta \leq 1;$$

$$\psi(q, \dot{q}) = q^1 - q + [\ln \eta + 1 - \eta - \frac{1}{2\eta}(\dot{q}^2 - (\eta - 1)^2)]\operatorname{sign} \dot{q}$$

$$\text{if} \quad \eta > 1 \quad \text{and} \quad |\dot{q}| < \eta - 1;$$

$$\psi(q, \dot{q}) = q^1 - q + [\ln(1 + |\dot{q}|) - |\dot{q}|]\operatorname{sign} \dot{q}$$

$$\text{if} \quad \eta > 1 \quad \text{and} \quad |\dot{q}| \geq \eta - 1 \tag{7.5.36}$$

The equation of the switching curve Γ in Figure 7.5.1 can be presented now as $\psi(q, \dot{q}) = 0$.

Equations 7.5.35 and 7.5.36 give the solution of the time-optimal control problem for the system governed by Equations 7.5.13 in dimensional variables (Equations 7.5.19). Substituting Equations 7.5.19 into Equations 7.5.35, 7.5.23, and 7.5.36, we can return to the original dimensional variables and parameters.

7.5.3. SUBOPTIMAL CONTROL OF MULTILINK ROBOTS

In Section 7.5.2 we obtained the feedback time-optimal control for an individual degree of freedom of the decoupled Equations 7.5.13. The optimal time of driving the whole system from the state specified by Equations 7.5.17 to the state in Equations 7.5.18 is obviously given by

$$T = \max_{1 \leq i \leq n} T_i \tag{7.5.37}$$

where T_i is the minimal time of bringing the ith link to the terminal state. The time T_i corresponds to the optimal control (Equations 7.5.35). The design of some robots, e.g., Cartesian-coordinate electromechanical manipulators, is such that their degrees of freedom are independent of each other. For such robots the control given by Equations 7.5.35 is exactly optimal. Here, the assumption that gear ratios are high is not necessary.

In the general case, the interdependence of different degrees of freedom takes place, but it is weak if $N_i \gg 1$. Therefore, we can apply the control given by Equations 7.5.35, based on the simplified model (Equations 7.5.13), to each degree of freedom and expect that the manipulator will be driven to the neighborhood of the terminal position (Equations 7.5.18) in the time close to Equation that in 7.5.37. Thus, this control can be regarded as suboptimal.

The positioning accuracy provided by such an approach can be sufficient for some transport robotic operations. If necessary, in a neighborhood of the terminal state the optimal control can be replaced by another robust control to correct the error caused by the dynamic interaction of different degrees of freedom and by inaccuracies in determining the parameters of the electro-mechanical system.

Let us give an illustrative example showing the results of the computer simulation of the robot motion under the suboptimal control proposed above. Consider a two-link manipulator moving in a horizontal plane. Note that many industrial robots have a two-link arm moving in a horizontal plane, in particular, SCARA-type robots. For this manipulator, the matrix A and vector S in the Lagrangian Equation 7.5.1 are given by

$$A = \left\| \begin{matrix} I_1 + I_2 + m_2 L_1^2 + 2m_2 L_1 L \cos q_2 & I_2 + m_2 L_1 L \cos q_2 \\ I_2 + m_2 L_1 L \cos q_2 & I_2 \end{matrix} \right\|$$

$$S = \left\| \begin{matrix} -m_2 L_1 L(2\dot{q}_1 + \dot{q}_2)\dot{q}_2 \sin q_2 \\ m_2 L_1 L \dot{q}_1^2 \sin q_2 \end{matrix} \right\| \tag{7.5.38}$$

Here, I_1 and I_2 are the inertia moments of the first and second links about the axes of the respective joints, m_2 is the mass of the second link, L_1 is the length of the first link, L is the distance between the joint connecting the links and the mass center of the second link.

Let us take the following values of the parameters in Equations 7.5.8 and 7.5.38:

$$I_1 = 21.8 \text{ kg·m}^2, \qquad I_2 = 5.9 \text{ kg·m}^2$$

$$J_1 = 9.1·10^{-4} \text{ kg·m}^2, \qquad J_2 = 2.45·10^{-4} \text{ kg·m}^2$$

$$m_2 = 58 \text{ kg}, \qquad L_1 \ 0.6 \text{ m}, \qquad L = 0.08 \text{ m}$$

$$N_1 = 180, \qquad N_2 = 163, \qquad R_1 = 3.68 \ \Omega, \qquad R_2 = 3.60 \ \Omega$$

$$h_1 = c_1 = 0.316 \text{ J/A}, \qquad h_2 = c_2 = 0.233 \text{ J/A},$$

$$b_1 = 0, \qquad b_2 = 0$$

$$U_1 = 110 \text{ V}, \qquad U_2 = 110 \text{ V}, \qquad j_1^0 = 3 \text{ A}, \qquad j_2^0 = 2 \text{ A} \tag{7.5.39}$$

These parameters correspond to the platform and arm of robot Universal-5 (see Sections 1.1.3, 5.4, and 7.4.2).

Let us introduce the dimensionless variables and parameters:

$$t' = \frac{U_2}{N_2 h_2} t, \qquad R_2' = \frac{U_2^2(I_i + J_i N_i^2)}{N_2^2 h_2^2 N_i c_i U_i} R_i$$

$$h_i' = \frac{N y i U_2}{N_2 h_2 U_i} h_i, \qquad c_i' = \frac{N y i U_2}{N_2 h_2 U_i} c_i, \qquad u_i' = \frac{u_i}{U_i}$$

$$I_1 = I_1 + I_2 + m_2 L_1^2, \qquad I_2 = I_2, \qquad i = 1, 2 \tag{7.5.40}$$

Substituting Equations 7.5.38 through 7.5.40 into Equations 7.5.8, we represent the latter equations in the dimensionless form as follows:

$$0.39\ddot{q}_1 + 1.5\,\dot{q}_1 = u_1 - 0.03\,\ddot{q}_1\cos q_2 - (0.03 + 0.01\cos q_2)\ddot{q}_2$$

$$+ 0.01(2\dot{q}_1 + \dot{q}_2)\dot{q}_2\sin q_2$$

$$0.09\ddot{q}_2 + \dot{q}_2 = u_2 - (0.04 + 0.02\cos q_2)\ddot{q}_1 - 0.02\sin q_2\cdot\dot{q}_1^2 \qquad (7.5.41)$$

Here and henceforth the primes are omitted.

Following the general approach described in Section 7.5.1, we simplify Equations 7.5.41 by omitting all nonlinear and coupling terms in their right-hand sides. After that, the equations and constraints (Equations 7.5.15 and 7.5.16) take the form

$$0.39\,\ddot{q}_1 + 1.5\,\dot{q}_1 = u_1, \qquad 0.09\ddot{q}_2 + \dot{q}_2 = u_2$$

$$|u_1| \le 1, \qquad |u_2| \le 1$$

$$|u_1 - 1.5\,\dot{q}_1| \le \eta_1 = 0.1, \qquad |u_2 - \dot{q}_2| \le \eta_2 = 0.07$$

$$\eta_i = R_i j_i^0 U_i^{-1}, \qquad i = 1, 2 \qquad (7.5.42)$$

Equations 7.5.42 are similar to Equations 7.5.13. Hence, we can use the results of Section 7.5.2 to obtain the feedback optimal control for this system.

For the numerical simulation, we assume that the system is at the state of rest at the initial instant: $\dot{q}^0 = 0$ in Equations 7.5.17. This condition is typical for robots. Equations 7.5.42 are invariant to the shift of the coordinates q_i as well as to the transformation $q_i \to -q_i$, $u_i \to -u_i$. Therefore, within the framework of the simplified model, the time of driving the ith link to the terminal state depends only on $\Delta q_i = |q_i^0 - q_i^1|$, if different links are controlled independently. This conclusion agrees with the experiments described in Section 7.4.2.

In what follows, we present some results of simulation of the manipulator dynamics under the controls $u_i(q_i, \dot{q}_i)$, $i = 1, 2$, optimal for the simplified decoupled model (Equations 7.5.42). These controls can be readily obtained from Equations 7.5.35 and 7.5.36 by the appropriate normalization of variables (see Equations 7.5.19). Before that, the respective coefficients in Equations 7.5.42 and 7.5.13 must be matched.

We apply the controls described above both to the simplified (Equations 7.5.42) and complete (Equations 7.5.41) models. The motion of the simplified model under the optimal control is obtained by analytical integration of Equations 7.5.42, whereas the nonlinear model (Equations 7.5.41) is simulated numerically.

Some results are given in Figures 7.5.2 through 7.5.5. Solid lines in Figures 7.5.2 and 7.5.3 show the optimal times $T_i(\Delta q_i)$ vs. the rotation angle

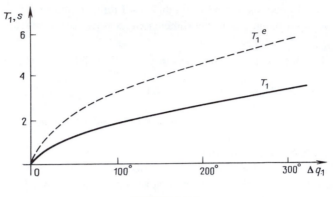

FIGURE 7.5.2.

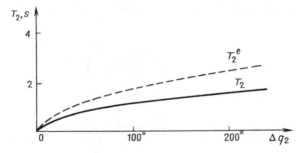

FIGURE 7.5.3.

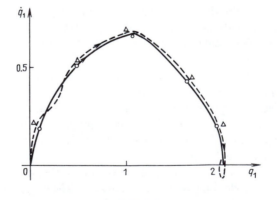

FIGURE 7.5.4.

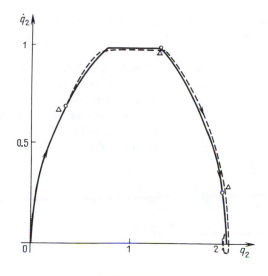

FIGURE 7.5.5.

for the platform ($i = 1$) and arm ($i = 2$) of robot Universal-5. These functions correspond to the simplified model (Equations 7.5.42) and are given by

$$T_i(\Delta q_i) = 2\left(\frac{B_i \Delta q_i}{c_i N_i j_i^0}\right)^{1/2} \quad \text{if} \quad \Delta q_i \leq \frac{B_i(U_i - R_i j_i^0)^2}{h_i^2 c_i N_i^3 j_i^0};$$

$$T_i(\Delta q_i) = \frac{B_i}{c_i N_i}\left(\frac{U_i - R_i j_i^0}{h_i N_i J_i^0} + \frac{R_i}{h_i N_i} \ln \frac{R_i j_i^0}{U_i - h_i N_i} + \frac{\omega}{j_i^0}\right)$$

$$\text{if} \quad \Delta q_i > \frac{B_i(U_i - R_i j_i^0)^2}{h_i^2 c_i N_i^3 j_i^0}, \qquad i = 1, 2;$$

$$B_1 = I_1 + I_2 + m_2 L_i^2 + J_1 N_1^2, \qquad B_2 = I_2 + J_2 N_2^2 \qquad (7.5.43)$$

Here, ω is the absolute value of the angular velocity at which the phase trajectory intersects the switching curve. To find ω, we are to solve the transcendental equation:

$$\Delta q_i = \frac{I_i}{2c_i N_i j_i^0}\left[\omega^2 + \left(\frac{U_i - R_i j_i^0}{h_i N_i}\right)^2\right]$$

$$+ \frac{I_i R_i U_i}{h_i^2 c_i N_i^3}\left(\ln \frac{R_i j_i^0}{U_i - h_i N_i \omega} + 1 - \frac{R_i j_i^0 + h_i N_i \omega}{U_i}\right) \qquad (7.5.44)$$

Equations 7.5.43 and 7.5.44 are valid, if $\eta_i = R_i j_i^0/U_i \leq 1$. This inequality holds in our example (see Equations 7.5.42.) For $\eta_i > 1$, similar expressions can be also given.

The dashed lines in Figures 7.5.2 and 7.5.3 show the experimental curves for the rotation times $T_i^e(\Delta q_i)$ of the platform ($i = 1$) and arm ($i = 2$) of our robot with the standard control unit (see Section 7.4.2).

To estimate the possible gain in the rotation time due to the control optimization, we introduce the absolute (ξ_i) and relative (ν_i) differences between the experimental time of rotation of the ith link ($T_i^e(\Delta q_i)$) and the predicted optimal time $T_i(\Delta q_i)$:

$$\xi_i(\Delta q_i) = T_i^e(\Delta q_i) - T_i(\Delta q_i), \qquad \nu_i(\Delta q_i) = \xi_i(\Delta q_i)/T_i^e(\Delta q_i)$$

The calculations show that both the absolute and relative differences are significant, namely,

$$\max_{0 \leq \Delta q_1 \leq s_1} \xi_1(\Delta q_1) = 2.39 \text{ s}, \qquad \max_{0 \leq \Delta q_2 \leq s_2} \xi_2(\Delta q_2) = 1 \text{ s}$$

$$\max_{0 \leq \Delta q_1 \leq s_1} \nu_1(\Delta q_1) = 0.39 \text{ s}, \qquad \max_{0 \leq \Delta q_2 \leq s_2} \nu_2(\Delta q_2) = 0.32$$

Here, $s_1 = 330°$ and $s_2 = 240°$ are the maximal admissible rotation angles for the platform and arm, respectively (see Equation 7.4.6).

Some results of simulation for the simplified (Equations 7.5.42) and complete nonlinear (Equations 7.5.41) models are presented in Figures 7.5.4 and 7.5.5, for the initial (Equations 7.5.17) and terminal (Equations 7.5.18) conditions given by

$$q_1^0 = q_2^0 = 0, \qquad \dot{q}_1^0 = \dot{q}_2^0 = 0 \qquad (7.5.45)$$

$$q_1^1 = q_2^1 = 2 \text{ rad}, \qquad \dot{q}_1^1 = \dot{q}_2^1 = 0 \qquad (7.5.46)$$

The solid and dashed lines corresponding to the simplified and complete models, respectively, show the projections of the phase trajectories of the manipulator onto the planes (q_1, $\dot{q}_1$) and (q_2, $\dot{q}_2$). The arrows indicate the direction of motion along the phase trajectories. The triangles Δ mark the consecutive states of the system described by Equations 7.5.41 at $t = \Delta t$, $2\Delta t$, $3\Delta t$, ..., where $\Delta t = 0.345$ s. The circles $\circ$ mark the respective states of the simplified system (Equations 7.5.42.)

For the simplified model (Equations 7.5.42), the minimal times of driving the platform (T_1) and arm (T_2) of the robot from the state presented by Equations 7.5.45 to the state presented by Equations 7.5.46 are

$$T_1 = 1.94 \text{ s}, \qquad T_2 = 1.13 \text{ s} \qquad (7.5.47)$$

These values are calculated by using Equations 7.5.43 and 7.5.44.

For the complete model (Equations 7.5.41), the angular velocities $\dot{q}_1$ and $\dot{q}_2$ become zero for the first time at

$$T_1' = 1.96 \text{ s}, \qquad T_2' = 1.16 \text{ s} \qquad (7.5.48)$$

respectively. At these instants, the corresponding rotation angles are

$$q_1(T_1') = 2.007 \text{ rad}, \qquad q_2(T_2') = 2.001 \text{ rad} \qquad (7.5.49)$$

The proximity of the solid and dashed lines in Figures 7.5.4 and 7.5.5, as well as the comparison of the numbers in Equations 7.5.46 through 7.5.49 indicate that the proposed feedback control brings the system to the desired state with a comparatively high accuracy. The time of motion is close to the optimal time calculated for the simplified system (Equation 7.5.42.) It gives ground to regard our decoupled control (Equations 7.5.35) as the time-suboptimal control for the complete nonlinear system described by Equations 7.5.41. The obtained accuracy of the manipulator positioning can be sufficient for "rough" transport operations, such as delivering workpieces from storage to the punching machine or packing the manufactured articles into a box. However, the majority of technological operations, e.g., assembly or spot welding, require high accuracy. In these cases, a modification of the control law is necessary which envisages a correction of the positioning errors in a vicinity of the terminal state. One of the possible ways for such correction is described, for instance, in Reference 18, where the combination of the suboptimal control and the linear regulator is proposed.

Chapter 8

APPLICATIONS OF ROBOTS

8.1. ROBOTIC SYSTEMS FOR MEASUREMENTS AND INSPECTION (RSMI)

In this chapter, we outline some new applications of robots which we practically dealt with at the Institute for Problems in Mechanics of the Russian Academy of Sciences. These applications are mostly connected with measurement and inspection operations. For such operations robots must be equipped with additional sensors and special data-processing systems. First, we briefly describe robotic systems for measurements and inspection (RSMI). After that, we consider pneumatic and optical sensors suitable for RSMI and discuss some kind of special grippers with the built-in sensors. Finally, we describe wall-climbing robots developed at the Institute for Problems in Mechanics.

It is hardly possible to create flexible and computer integrated manufacturing systems without automation of inspection, measurement, and diagnostic operations. It seems promising to use manipulation robots for such operations, thus integrating inspection and manipulation functions in a special RSMI. This combination allows us to reduce considerably the time expenditure. RSMI can perform inspection operations either parallel with technological operations on various stages of the manufacturing process, or on the final stage, for inspecting manufactured articles. In some cases, technological and measurement operations can be carried out by the same robot. At present, RSMI are used for assembly, welding, sorting, inspection of complex surfaces (for instance, turbine blades), nondestructive testing of materials, and other operations and processes which involve measurements as their essential parts. For the description of measurement robotic systems and their applications see, for example, References 97,120,177,183,208.

Compared with a usual manipulation robot, RSMI has a rather more developed information system. It is equipped with additional specialized sensors, transducers, and microcomputers for receiving and processing the measurement information. Sometimes, the sensors are built-in inside the robot gripper.

As a rule, the control system of RSMI contains two loops, namely, the motion programming and diagnostic ones. The former is responsible for the motion of the manipulator, whereas the latter controls the measurement operations, processes the measurement data, and classifies the inspected objects into groups according to the chosen parameters.

The motion programming loop, apart from usual sensors, can include some special transducers serving, e.g., for detecting the edge of an inspected object or tracking its surface. This loop is also responsible for the adaptation of the robot to environmental changes. The diagnostic loop includes the sensors and transducers designed specially for detecting different imperfections (cracks, cavities, etc.) or recognizing qualitative features of objects (color, surface roughness, shape, etc.).

It is a distinctive feature of RSMI that they can act as automatic researchers. This makes it promising to use such systems for investigation work in an environment that is hazardous or inaccessible to man (for example, in space, on the ocean bottom, or inside nuclear reactors).

8.2. SENSORS FOR RSMI

In Chapter 1, we gave general information about the transducers of different types which are used in manipulation robots. In this section we will discuss the application of optical and pneumatic sensors in robotic systems for measurements and inspection.

Optical sensors can be placed either outside or on the robot. Sometimes, optical transducers are located at the robot gripper. The transducers of the following two types are mostly used: solid photosensitive matrices and TV camera tubes (vidicons or image dissectors). The artificial vision systems based on TV camera tubes have high resolution and sensitivity. However, their dimensions and mass are comparatively large. To reduce the dimensions of the primary receiver located at the gripper, optical fibers are used for transmitting the light from the primary receiver to the transducer placed outside the robot. It seems to be most promising to use solid photosensitive matrices in RSMI. Such matrices possess good optical characteristics and, at the same time, they are compact and light.

Artificial vision systems can receive and process rapidly large amounts of information. This property makes these systems adequate to the task of RSMI. The block diagram of the typical artificial vision system for an inspection robot is given in Figure 8.2.1. The system contains the light source (1), TV cameras (2), the optical unit (3) for forming the image and converting it into the video signal, the interface unit (4), the computer (5), the monitor (6), and the block of on-off switches (7). The optical parameters of the inspected object are measured by means of different photometric devices based on such physical phenomena as interference, diffraction, polarization, and dispersion of light, as well as on nonlinear optical effects of the interaction of the laser radiation with media. For one-dimensional measurements, optical rulers are used; two-dimensional images are obtained by means of photomatrices; for generating three-dimensional images, multilayered matrices, two or three TV cameras, or holographic devices are used.

In case the inspected object is moving, it is sometimes necessary to synchronize the formation and processing of the image. Synchronization can be implemented, for instance, by using a stroboscopic light source or a shutter in the TV camera.

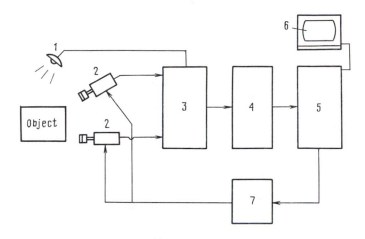

FIGURE 8.2.1.

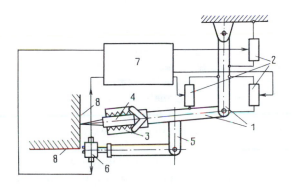

FIGURE 8.2.2.

Consider now some pneumatic sensors suitable for inspection robots.[104,122] In Figure 8.2.2, the tracking arm is shown that follows the surface of the inspected object. The distance between the gripper and the surface is measured by the pneumatic jet sensor located at the gripper. The arm includes the links (1), the drives (2), the gripper (3) with the testing head (4), the holder (5), the pneumatic sensor (6), and the control unit (7). During the inspection, the sensor (6) controls the position of the testing head (4) with respect to the inspected surface (8). In case the distance between the surface and the sensor changes or the sensor passes the edge of the inspected object, the pneumatic signals are generated and applied to the control unit (7). The control unit processes the pneumatic signals and, if necessary, generates the control commands for the drives to correct deviations from working conditions.

The pneumatic sensor (Figure 8.2.3) consists of the body (1), the measurement channels (2), the input (3) and output (4) air restrictors, the supply

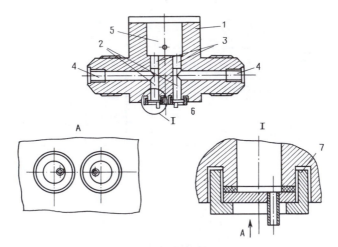

FIGURE 8.2.3.

reservoir (5), the jet nozzles (6), and the manual regulator of the distance
between the nozzles (7). The sensor works in the following way. The compressed air passes through the filter and pressure regulator, and then is fed
to the supply reservoir (5). Passing through the input air restrictors (3), the
air goes to the measurement channels (2). The pressures in these channels
depend on the positions of the jet nozzles (6) with respect to the inspected
surface. If the nozzle of the channel faces the workpiece surface, the pressure
in the channel clearly reflects the distance between the nozzle and the surface.
In case the nozzle does not face the surface, the pressure in the corresponding
channel is much smaller. Hence, if the sensor tracks the edge of the workpiece
in such a way that only one nozzle faces the workpiece surface, then the
difference between the pressures indicates that the gripper is near the edge.
The accuracy of the sensor depends on the distance between the nozzles. This
distance can be adjusted by the regulator (7). From the measurements channels, the air goes through the output restrictors (4) to the control unit, which
generates commands for the drives of the manipulator arm.

Figure 8.2.4 presents the pressure p_2 at the output port of the sensor vs.
the distance s from the surface to the nozzle of the corresponding channel,
for different supply pressures p_1. The curves 1, 2, and 3 correspond to $p_1 =$
0.5 MPa, $p_1 = 0.2$ MPa, and $p_1 = 0.05$ MPa, respectively. We can see
from the graphs that both the sensitivity and range of operational distances
of the sensor increase as the supply pressure grows.

If the inspected object is of a complex shape, it is appropriate to use
fluidic acoustical transducers which contain Helmholtz resonators with built-
in restrictors (Figure 8.2.5). These transducers serve for detecting obstacles
supplied with sound emitters. The work of such transducers is based on the
destruction of the laminar air jet flowing from the input restrictor (1) through

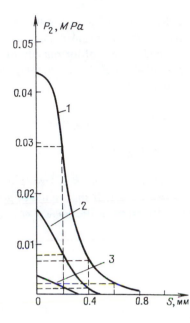

FIGURE 8.2.4.

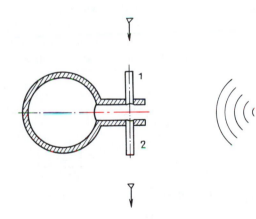

FIGURE 8.2.5.

the cavity to the output restrictor (2). The destruction occurs if the pressure in the resonant cavity exceeds certain critical values.

The Helmholtz resonator is a multifrequency oscillatory system with the lowest resonant frequency equal to

$$\omega_0 = a(\pi R^2 V^{-1} l^{-1})^{1/2}$$
$$a = (\gamma p_0 \rho_0^{-1})^{1/2}$$

Here, R is the radius of the resonator tube, l is the length of the tube, V is the volume of the resonant cavity, $\gamma = 1.4$ is the adiabatic exponent of air, p_0 and ρ_0 are the pressure and density of air outside the resonator, and a is the sound velocity. Free oscillations of the air pressure in the resonant cavity are governed by the equation,

$$\ddot{p} + r\dot{p} + \omega_0^2 p = 0$$

$$r = 8\mu\rho^{-1}R^{-2} \tag{8.2.1}$$

where p is the difference between the pressure inside and outside the cavity, and μ is the dynamic viscosity of air. If the oscillator governed by Equation 8.2.1 is excited harmonically with the frequency ω_0, the amplitude of the forced oscillations is given by

$$p_c = \omega_0 r^{-1} p_i$$

$$\tag{8.2.2}$$

Here, p_c and p_i are the amplitudes of the pressure oscillations in the cavity and at the input port, respectively. Harmonic oscillations at the input port are generated by special acoustic radiators placed on the inspected object surface. Note that if $\omega_0 \gg r$ we have $p_c \gg p_i$ and, hence, the Helmholtz resonator is highly sensitive to external oscillations with the resonant frequency ω_0.

It is shown[80] that the pressure p_L of the submersed laminar jet (flowing out of the input restrictor (1)) at the input port of the output restrictor (2) can be calculated as follows:

$$p_L = \frac{\alpha_1 \rho}{2(1 + \alpha_2 \alpha_0^2)} \left(\frac{A}{1 + Br_2^2} \right)^2$$

$$\alpha_0 = \left\{ \frac{r_2}{r_l n^2} \left[1 + \xi\left(1 - \frac{r_l^2}{r_2^2} \right) \right] - \alpha_2 \right\}^{-1/2}$$

$$\alpha_1 = [(1 + Br_2^2)^3 - (1 + Br_2^2)^{-2}](5Br_2^2)^{-1}$$

$$\alpha_2 = 1 + (\alpha_1 - 1)(r_l/r_0)$$

$$A = \frac{Q^2 \rho}{2\pi^2 \mu x r_1^2}$$

$$B = \left(\frac{Q\rho}{4\pi\mu x r_1} \right)^2$$

Here, ξ is the resistance coefficient of the restrictor (2) (ξ ranges from 0.5 to 0.7); Q is the rate of the air flow through the restrictor 1; r_1 and r_2 are the radii of the restrictors 1 and 2, respectively, n is the number of load restrictors

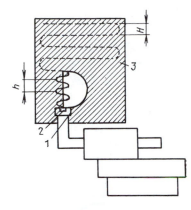

FIGURE 8.3.1.

connected to the output port of the restrictor 2; r_l is the radius of the load restrictors; and x is the distance between the output nozzle of the restrictor 1 and the input nozzle of the restrictor 2. As the sensor approaches the surface with acoustic radiators on it, the pressure p_i in Equation 8.2.2 grows, and the pressure p_c in the resonant cavity becomes greater than p_L. This causes the destruction of the laminar jet, the pressure at the output port of the restrictor 2 drops, and the sensor registers the obstacle.

8.3. SENSORY GRIPPERS

For robots carrying out operations associated with measurements, it is often expedient to use grippers with built-in sensors. Such sensory grippers can significantly enhance the accuracy of performing the prescribed task, because they facilitate the fine positioning of the tool or another manipulation object. Very often, sensors built-in inside grippers are the pneumatic ones. This is explained by the relative simplicity of pneumatic transducers, and also by the fact that in many industrial robots the drives of grasping mechanisms and the pneumatic sensors use the same energy source (compressed air). Pneumatic sensors are most appropriate for measuring the geometrical parameters of objects and detecting imperfections associated with deviations of the geometrical parameters from their nominal values. Pneumatic sensory grippers can be used for measuring the dimensions of workpieces, detecting holes and cracks, determining the surface roughness, etc.

To inspect an object, the robot with a pneumatic sensory gripper can scan the surface of the object. In Figure 8.3.1, line-by-line scanning is shown which is aimed at the detection of a hole and the measurement of its diameter. Starting from the upper edge of the surface (1), the gripper (2) with the sensor scans the surface (3) with a prescribed step *(H)*. The size of the step depends on the geometrical parameters of the inspected object, e.g., on the minimal

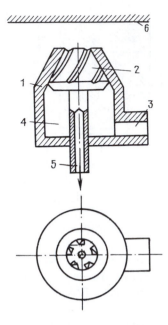

FIGURE 8.3.2.

expected diameter of a hole on its surface. When the sensor meets the edge
of a hole, the special signal is generated and fed to the control system.
Following the signal, the robot changes the mode of scanning. The step H is
replaced by a smaller one *(h < H),* and the range of scanning is also reduced
(see Figure 8.3.1). Thus, the robot can determine the hole boundary and track
it. During the tracking, the information about the hole location and shape is
transmitted to a computer for subsequent processing.

The sketch of a sensor for scanning is presented in Figure 8.3.2. The
sensor consists of the body (1) with spiral channels (2), the input channel
(3), the chamber (4), and the measurement channel (5). When the vortex flux
of air flowing out of the spiral channels hits the surface (6) of the inspected
object, the pressure in the measurement channel changes, thus indicating the
proximity of the surface.

In Figure 8.3.3, the vacuum gripper with the sensor for vacuum control
is shown. The gripper consists of the input channel (1), the ejector (2), the
vacuum channel (3), the vacuum suckers (4), and the sensor (5). The sensor
(5) consists of the vacuum chamber (6), the membrane (7), the rigid central
part (8), the adjusting screw (9), the spring (10), and the relay (11). The
sensor works as follows. If the contact between the surface of a workpiece
and the suckers (4) is reliable, air in the vacuum channel (3) and the chamber
(6) becomes rare, and the membrane (7) bends towards the relay contact
leaves. When the pressure in the vacuum chamber becomes less than a certain

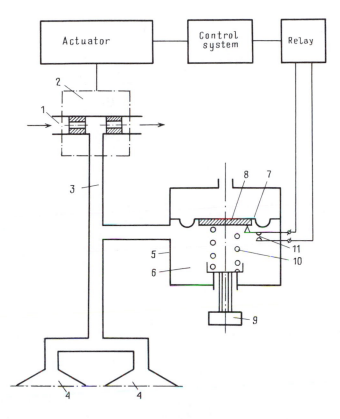

FIGURE 8.3.3.

value indicating that the workpiece is reliably taken by the gripper, the central part (8) of the membrane (7) connects the relay contact points, and the signal is applied to the control unit. The initial position of the membrane should correspond to the degree of vacuum required for reliable contact with the object and can be adjusted by means of the screw (9).

Figure 8.3.4 presents the sketch of the measuring gripper that can determine the external and internal dimensions of objects placed in it (for example, the external and internal diameters of rings and tubes). The gripper includes the lever mechanism (1), the adjustable holder (2), the pneumatic sensor (3), the jaws (4), and the plate (5) attached to one of the levers rigidly connected with the jaws. For measuring internal dimensions, the gripper is provided with accessory jaws (shown by the dashed line in Figure 8.3.4).

The distance between the jaws (equal to the size of the grasped object) determines the gap between the jet nozzle and the plate (5). By measuring the gap, the sensor (3) defines the size of the object. The normal gap (corresponding to the normal size of the inspected object) is adjusted by rotating the holder (2).

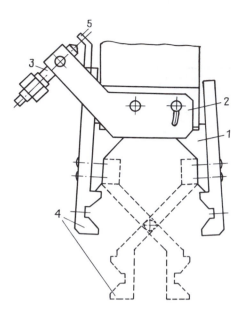

FIGURE 8.3.4.

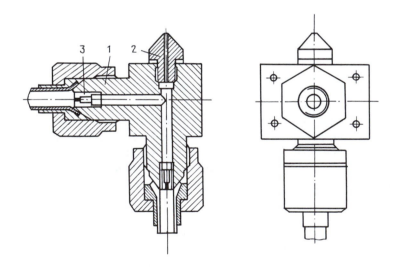

FIGURE 8.3.5.

The sensor is shown in Figure 8.3.5. It includes the body (1), the nozzle (2), and the input restrictor (3). The sensor is included into the circuit containing a closed measurement chamber which communicates with a pneumatic transducer. The sensor is provided with a set of accessory fluidic resistors. By choosing the resistors, one can achieve the required accuracy and sensitivity of measurements.

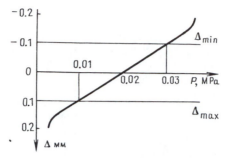

FIGURE 8.3.6.

The experimental characteristic of the sensor presented in Figure 8.3.6 shows the deviation Δ of the size of the inspected object from the nominal size vs. the pressure p in the information channel. The sensor measures deviations within the range ± 0.1 mm (in this range the characteristic in Figure 8.3.6 is linear) with an accuracy up to several micrometers.

The described gripper allows combining transport and measurement operations and thus reduces considerably the time of sizing workpieces or manufactured articles.

The measuring grippers equipped with pneumatic sensors developed at the Institute for Problems in Mechanics are shown in Figures 8.3.7 and 8.3.8.

8.4. WALL-CLIMBING ROBOTS (WCR)

In this section we present a brief outline of a new type of robotic systems, namely, wall-climbing robots (WCR), focusing mostly on the developments made at the Institute for Problems in Mechanics of the Russian Academy of Sciences. In many applications, WCR are equipped with manipulators which carry out technological, inspection, or other operations. Therefore, scientific and engineering problems related to WCR and manipulation robots are closely connected.

WCR are new automatic machines that can move along vertical, sloping, and ceiling surfaces, accomplishing different tasks. WCR can be used for the construction and maintenance of buildings and structures, for cleaning, painting, welding, cutting, inspection, etc. They can work under extreme conditions, in hazardous and unstructured environments.

WCR are complex systems that include mechanical, pneumatic, electric, and electronic parts controlled by a man and/or a computer. The design of these robots is based on the results of scientific research in mechanics and automatic control and employs modern achievements in high technologies. WCR can also be regarded from the viewpoint of bionics because they carry out motions along vertical and ceiling surfaces; only some insects can perform such motions.

FIGURE 8.3.7.

Research and development of WCR are conducted in Japan, U.S., U.K., Russia, and other countries. Several models of WCR were developed and used for different practical purposes. Some results in this field are reflected in References 74,88,90,116,176,220.

WCR can be regarded as a special kind of automatic walking machine (see, e.g., Reference 182). The distinctive feature of WCR is the presence of special grippers on their legs for reliable contact of the machine with the vertical surface. At present, two types of grippers are mostly developed and used. These are magnetic grippers (for ferromagnetic surfaces) and vacuum suckers. In both cases, the grippers are strongly pressed against the surface, and dry friction between the grippers and the surface prevents the robot from slipping down. Therefore, to determine the parameters (e.g., the contact area and the air rarefaction inside the vacuum sucker) and number of the grippers

FIGURE 8.3.8.

for a concrete WCR, as well as to control the motion of the robot, one must rely on the physical laws governing the behavior of a body on a rough surface.

At the Institute for Problems in Mechanics of the Russian Academy of Sciences, WCR have been under development since 1984. The general principles of the design of WCR were elaborated. WCR (see Figure 8.4.1) consist of transport, technological, and control modules.

The transport module includes two platforms (1 and 2) that can move with respect to each other. On the platforms, ejector-type vacuum grippers (3) are mounted which ensure reliable contact with the wall.

The motion of the robot in the prescribed direction occurs as follows. At any instant, one of the platforms, say 1, is attached to the wall by means of its grippers, while the grippers of the platform 2 are lifted and do not touch the wall. The platform 2 is moved with respect to the platform 1 by means of pneumatic drives. Then the platform 2 stops, its grippers approach the wall, and are sucked onto it due to vacuum created between the grippers and the wall. After that, the vacuum in the grippers of the platform 1 is removed,

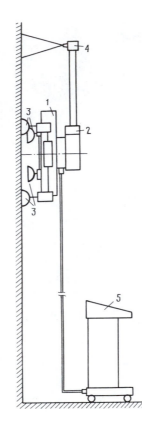

FIGURE 8.4.1.

these grippers are disconnected from the wall and moved closer to the platform. Now platform 1 moves with respect to platform 2, which is fixed to the wall, and the process is repeated. A special mechanism makes it possible to change the direction of motion. On some models of WCR, the rotation of the platform about the axis perpendicular to the wall is also envisaged.

The compressed air for pneumatic drives and grippers is supplied to the WCR through the air hose connecting the robot with the source of compressed air. A compressor, usually located on the ground, can be replaced by a cylinder with compressed air mounted on the robot.

To enhance the reliability of contact of the WCR with the wall, the ejector-type vacuum grippers of a special design were developed. The gripper consists of the central load-bearing metallic part, with small pins on the contact surface, and a special cover made of a soft elastic material. The edge of the cover closely fits the wall. This does not permit atmosphere air to flow inside the sucker and thus maintains vacuum.

The technological module (4), depending on the task, can include manipulators, sensors, and special equipment. WCR developed at the Institute

FIGURE 8.4.2.

for Problems in Mechanics can carry out such technological operations as cleaning by means of special brushes and vacuum cleaners, painting, cutting, welding, and inspection using different sensors. For example, they can carry TV camera and ultrasonic sensors. The WCR can also wash windows, check the quality of welds on structures, take part in different fire-fighting measures, and perform many other functions. WCR with different technological modules are shown in Figures 8.4.2 and 8.4.3. Figure 8.4.2 shows the WCR equipped with a module for cutting, while Figure 8.4.3 shows the robot for ultrasonic diagnostics.

The control module (5) is placed on the ground and is connected with the robot by a cable. The cable can be easily replaced by a wireless connection. The control unit receives information from various sensors placed on the robot. This information indicates the positions of different parts of the robot, pressures, forces, etc. The signals from the technological equipment are also applied to the control module. This module includes a controller and a personal computer (PC); an operator can also be involved in the control process. The control unit is supplied with special algorithms and programs that process the data obtained from the sensors and implement the feedback control both of the transport and of the technological modules.

Scientific research, both theoretical and experimental, was performed that helped to design WCR and choose their parameters.[1,42,61,62,99,100,105–109]

In References 61 and 62, the equilibrium of a rigid body on a rough plane under the action of external forces, normal surface reactions, and dry (Coulomb) friction is considered. Guaranteed equilibrium conditions are obtained that remove the static uncertainty inherent in the general problem of the equilibrium of a rigid body on a rigid surface.

In Reference 1, the guaranteed equilibrium conditions are applied for the estimation of the load-carrying capacity of the WCR with vacuum grippers.

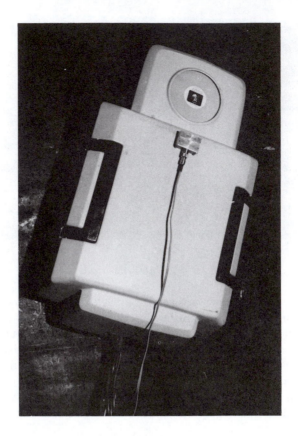

FIGURE 8.4.3.

Besides, the equilibrium conditions for the WCR are obtained which take into account the elastic compliance of the robot legs. Both approaches yield close numerical estimates of the load-carrying capacity of the robot.

WCR with two-link legs and vacuum grippers is considered in Reference 42. To ensure the equilibrium of such a robot on a vertical surface, the static control torques must be applied to the leg joints. The equilibrium conditions are obtained, and the following optimization problem is treated: how to place the grippers on the surface and distribute the control torques over the joints so as to achieve the equilibrium of the robot under minimal air rarefaction inside the vacuum suckers.

Motions of WCR under external disturbances were studied by Gradetsky and Akselrod.[99] The behavior of vacuum grippers in the presence of vibrations was investigated by Gradetsky et al.[100]

At the Institute for Problems in Mechanics, several models of WCR for different purposes were developed. They can carry a payload of up to 100 kg and move at a speed of up to 1 m/min. Special tests showed that the robots

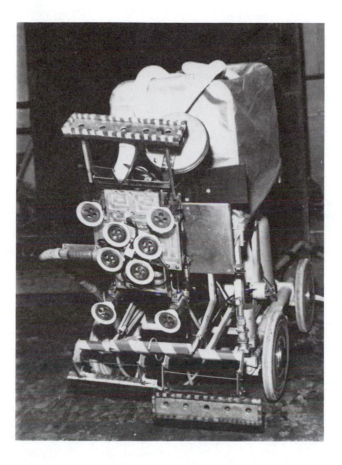

FIGURE 8.4.4.

can operate on wet and hot (at temperatures up to 300°C) surfaces, and also in the presence of vibrations. These properties are important for applications in fire fighting and other extreme conditions. Fire-fighting WCR developed at the Institute are being produced in industry.

Two WCR developed at the Institute participated in the First World Robot Olympics (Glasgow, 1990) and won two medals.

WCR can be a part of more complex systems. For example, the robotic system for cleaning the internal walls of nuclear power plants was created at the Institute for Problems in Mechanics. The system includes a mobile wheeled robot for horizontal motions, and a WCR (see Figure 8.4.4).

REFERENCES

1. **Abarinov, A. V., Akselrod, B. V., Bolotnik, N. N., Veshnikov, V. B., Gomozov, A. V., Gradetskiy, V. G., Zinovyev, F. V., Meshman, L. M., Moskalev, V. S., Rachkov, M. Yu., and Chernousko, F. L.,** A robot system for moving over vertical surfaces, *Sov. J. Comput. Syst. Sci.,* 27(3), 130, 1989.

2. **Ailon, A. and Langholz, G.,** On the existence of time-optimal control of mechanical manipulators, *J. Optimization Theor. Appl.,* 46, 1, 1985.

3. **Akulenko, L. D.,** *Problems and Methods of Optimal Control,* Kluwer Academic Publishers, Dordrecht, 1993 in press.

4. **Akulenko, L. D. and Bolotnik, N. N.,** On controlled rotation of an elastic rod, *J. Appl. Math. Mech. (USSR),* 46, 465, 1982.

5. **Akulenko, L. D. and Bolotnik, N. N.,** On control of rotation of an elastic manipulator link, *Izv. Akad. Nauk SSSR. Tekh. Kibern.,* 1, 167, 1984 (in Russian).

6. **Akulenko, L. D. and Bolotnik, N. N.,** Synthesis of optimal control of transport motions of manipulation robots, *Mech. Solids (USSR),* 21(4), 18, 1986.

7. **Akulenko, L. D., Bolotnik, N. N., Chernousko, F. L., and Kaplunov, A. A.,** Optimal control of manipulation robots, in *Proc. Ninth Triennnial World Congress of IFAC,* Vol. 1, Gertler, J. and Keviczky, L., Eds., Pergamon Press, Oxford, 1985, 331.

8. **Akulenko, L. D., Bolotnik, N. N., and Kaplunov, A. A.,** Some control modes for industrial manipulators, *Izv. Akad. Nauk SSSR. Tekh. Kibern.,* 6, 44, 1985.

9. **Akulenko, L. D. and Gukasyan, A. A.,** Control of plane motions of an elastic manipulator element, *Mech. Solids (USSR),* 18(5), 31, 1983.

10. **Akulenko, L. D., Mikhailov, S. A., and Chernousko, F. L.,** Simulation of the dynamics of a manipulator with elastic elements, *Mech. Solids (USSR),* 16(3), 111, 1981.

11. **Akulenko, L. D. and Mikhailov, S. A.,** Synthesizing the control of the rotations of an elastic link in an electromechanical manipulator robot, *Sov. J. Comput. Syst. Sci.,* 26(6), 99, 1988.

12. **Alberts, Th. E., Dickerson, St., and Book, W. J.,** Modelling and control of flexible manipulators, in *Proc. Robots 9 Conf.,* Detroit, 1985, 1/1.

13. **Ardayfio, D. D.,** Model formulation of flexible manipulator arms, in *Math. Model. Sci. Technol., 4th Int. Conf.,* New York, 1983.

14. **Asada, H., Ma, Z.-D., and Tokumaru, H.,** Inverse dynamics of flexible robot arms: modelling and computation for trajectory control, *Trans. ASME. J. Dyn. Syst. Meas. Control,* 112, 177, 1990.

15. **Asada, H. and Stotine, J.-J. E.,** *Robot Analysis and Control,* John Wiley & Sons, New York, 1985.

16. **Avetisyan, V. V.,** Optimization of transport motions of manipulation robots with limited head-shedding capacity, *Sov. J. Comput. Syst. Sci.,* 26(3), 45, 1988.

17. **Avetisyan, V. V., Akulenko, L. D., and Bolotnik, N. N.,** Optimization of control modes of manipulation robots with regard of the energy consumption, *Izv. Akad. Nauk SSSR, Tekh. Kibern.,* 3, 100, 1987 (in Russian).

18. **Avetisyan, V. V. and Bolotnik, N. N.,** Suboptimal control of an electromechanical manipulator with a high degree of positioning accuracy, *Mech. Solids (USSR),* 25(5), 32, 1990.

19. **Avetisyan, V. V., Bolotnik, N. N., and Chernousko, F. L.,** Optimal programmed motion of a two-link manipulator, *Sov. J. Comput. Syst. Sci.,* 23(5), 65, 1985.

20. **Balas, M. J.,** Feedback control of flexible systems, *IEEE Trans. Autom. Control,* AC-23, 673, 1978.

21. **Bayo, E., Movaghar, R., and Medus, M.,** Inverse dynamics of a single-link flexible robot. Analytical and experimental results, *Int. J. Robotics Autom.,* 3, 150, 1988.

22. **Bayo, E., Papadopoulos, P., Stubbe, J., and Serna, M. A.,** Inverse dynamics and kinematics of multilink elastic robots: an iterative frequency domain approach, *Int. J. Robotics Res.,* 8, 49, 1989.

23. **Becker, P.-J., Jacubasch, A., and Kuntze, H.-B.,** On the design of a computer controlled elastic industrial robot, in *ACI 83. Appl. Control and Identification:* Proc. of IASTED Symposium, 1, Copenhagen, 1983.

24. **Belyanin, P. N.,** *Industrial Robots and their Applications,* Mashinostroyeniye, Moscow, 1983 (in Russian).

25. **Berbyuk, V. E.,** *Dynamics and Optimization of Robotic Systems,* Naukova Dumka, Kiev, 1989 (in Russian).

26. **Berbyuk, V. E. and Demidyuk, M. V.,** Controlled motion of an elastic manipulator with distributed parameters, *Mech. Solids (USSR),* 19(2), 57, 1984.

27. **Berbyuk, V. E. and Demidyuk, M. V.,** Parametric optimization in problems of dynamics and control of motion of an elastic manipulator with distributed parameters, *Mech. Solids (USSR),* 21(2), 78, 1986.

28. **Berbyuk, V. E., Demidyuk, M. V., and Ivakh, G. F.,** Problems of optimization of structures and control laws of electro-mechanical manipulators, *Izv. Akad. Nauk SSSR, Tekh. Kibern.,* 3, 113, 1987 (in Russian).

29. **Berbyuk, V. E. and Ivakh, G. F.,** Mathematical design of gantry robots for automating assembly processes, *Sov. J. Comput. Syst. Sci.,* 27(2), 1, 1989.

30. **Berbyuk, V. E. and Yanchak, Ya. I.,** Minimum-time optimization of transport motions of a gantry robot, *Izv. Akad. Nauk SSSR, Tekh. Kibern.,* 1, 126, 1991 (in Russian).

31. **Bobrow, J. E.,** Optimal robot path planning using the minimum-time criterion, *IEEE J. Robotics Autom.,* 4, 443, 1988.

32. **Bobrow, J. E., Dubowsky, S., and Gibson, J. S.,** On the optimal control of robotic manipulators with actuators constraints, in *Proc. 1983 Am. Control Conf.,* San Francisco, 1983.

33. **Bobrow, J. E., Dubowsky, S., and Gibson, J. S.,** Time-optimal control of robotic manipulators along specified paths, *Int. J. Robotics Res.,* 4, 3, 1985.

34. **Bolotnik, N. N. and Chernousko, F. L.,** Optimization of manipulation robot control, *Sov. J. Comput. Syst. Sci.,* 28(5), 127, 1990.

35. **Bolotnik, N. N. and Dobrynina, I. S.,** Control of rotation of an elastic manipulator link with damping of finite number of oscillation modes, *Izv. Akad. Nauk SSSR, Tekh. Kibern.,* 4, 207, 1989 (in Russian).

36. **Bolotnik, N. N., Gorbachev, N. V., and Shukhov, A. G.,** Combined nearly-optimal control of an electromechanical system, *Izv. Akad. Nauk SSSR, Tekh. Kibern.,* 6, 192, 1991 (in Russian).

37. **Bolotnik, N. N., Gorbachev, N. V., and Shukhov, A. G.,** Optimization of control of an electromechanical system with respect to the minimax performance index, *Izv. Akad. Nauk SSSR, Mekh. Tverd. Tela,* 6, 30, 1992 (in Russian).

38. **Bolotnik, N. N. and Gukasyan, A. A.,** Control of motion of a manipulator with allowance for elastic vibrations of the boom, *Mech. Solids (USSR),* 19(4), 34, 1984.

39. **Bolotnik, N. N. and Kaplunov, A. A.,** Optimal rectilinear transfer of a load by means of a two-link manipulator, *Izv. Akad. Nauk SSSR, Tekh. Kibern.,* 1, 160, 1982 (in Russian).

40. **Bolotnik, N. N. and Kaplunov, A. A.,** Optimization of control and configurations of a two-link manipulator, *Izv. Akad. Nauk SSSR, Tekh. Kibern.,* 4, 144, 1983 (in Russian).

41. **Bolotnik, N. N. and Kaplunov, A. A.,** Synthesis of control of a two-element manipulator, *Mech. Solids (USSR),* 20(3), 57, 1985.

42. **Bolotnik, N. N. and Nandi, G. C.,** On the control of equilibrium of a wall climbing robot, *Izv. Akad. Nauk SSSR, Mekh. Tverd. Tela,* 4, 58, 1992 (in Russian).

43. **Book, W. J.,** Analysis of massless elastic chains with servo controlled joints, *Trans. ASME. J. Dyn. Syst. Meas. Control,* 101, 187, 1979.

44. **Book, W. J., Maizza-Neto, O., and Whitney, D. E.,** Feedback control of two beam, two joint systems with distributed flexibility, *Trans. ASME J. Dyn. Syst. Meas. Control,* 97, 424, 1975.

45. **Book, W. and Majette, M.,** Controller design for flexible distributed parameter mechanical arms via combined frequency domain techniques, *Trans. ASME. J. Dyn. Syst. Meas. Control,* 105, 245, 1983.

46. **Borisov, V. F. and Zelikin, M. I.,** Modes with switchings of increasing frequency in the problem of controlling a robot, *J. Appl. Math. Mech. (USSR),* 52, 731, 1988.

47. **Bryson, A. E. and Ho, Y. C.,** *Applied Optimal Control,* John Wiley & Sons, New York, 1975.

48. **Chelpanov, I. B. and Kolpashnikov, S. N.,** Grippers for Industrial Robots, *Mashinostroyeniye,* Leningrad, 1989 (in Russian).

49. **Chen, Y.,** On the structure of the time-optimal controls for robotic manipulators, *IEEE Trans. Autom. Control,* AC-34, 115, 1989.

50. **Chen. Y.,** Existence and structure of minimum-time control for multiple robot arm handling a common object, *Int. J. Control,* 53, 855, 1991.

51. **Chen, Y. P.,** Solving robot trajectory planning problems with uniform cubic B-splines, *Optimal Control Appl. Methods,* 12, 247, 1991.

52. **Chen, Y. P. and Yeung, K. S.,** Sliding-mode control of multi-link flexible manipulators, *Int. J. Control,* 54, 257, 1991.

53. **Chernousko, F. L.,** Motion of a rigid body with cavities filled with viscous fluid under small Reynolds numbers, *J. Vychisl. Mat. Mat. Fiz.,* 5, 1049, 1965 (in Russian).

54. **Chernousko, F. L.,** *Motion of a Rigid Body with Cavities Containing Viscous Fluid,* Computing Center of the USSR Academy of Science, Moscow, 1968 (in Russian).

55. **Chernousko, F. L.,** Motion of a rigid body with moving internal masses, *Mech. Solids (USSR),* 8(4), 27, 1973.

56. **Chernousko, F. L.,** On the motion of a solid body with elastic and dissipative elements, *J. Appl. Math. Mech. (USSR),* 42, 32, 1978.

57. **Chernousko, F. L.,** Dynamics of controlled motion of elastic manipulator, *Izv. Akad. Nauk SSSR, Tekh. Kibern.,* 5, 142, 1981 (in Russian).

58. **Chernousko, F. L.,** Dynamics of systems with elastic elements of large stiffness, *Mech. Solids (USSR),* 18(4), 99, 1983.

59. **Chernousko, F. L.,** Dynamics of robots with elastic elements, in *Symp. Grundlagen der Dynamik und Steuerung von Industrie Robotern,* Vorträge, Band 1, Heimann, B., Ed., Berlin, 1985, 12.

60. **Chernousko, F. L.,** Asymptotic methods in dynamics of systems with elastic and dissipative elements, in *Proc. 4th Int. Conf. on Boundary and Interior Layers,* Godunov, S. K., Miller, J. J. H., and Novikov, V. A., Eds., Boole, Dublin, 1987, 15.

61. **Chernousko, F. L.,** Equilibrium conditions for a solid on a rough plane, *Mech. Solids (USSR),* 23(6), 1, 1988.

62. **Chernousko, F. L.,** On the mechanics of a climbing robot, *Mech. Syst. Eng.,* 1, 219, 1990.

63. **Chernousko, F. L.,** Decomposition and suboptimal control in dynamical systems, *J. Appl. Math. Mech. (USSR),* 54, 727, 1990.

64. **Chernousko, F. L.,** Asymptotic analysis for dynamics of rigid body containing elastic elements and viscous fluid, in *Dynamical Problems of Rigid-Elastic Systems and Structures, IUTAM Symp.,* Banichuk, N. V., Klimov, D. M., and Schiehlen, W., Eds., Springer-Verlag, Berlin, 1991, 55.

65. **Chernousko, F. L.,** Decomposition and synthesis of control in dynamical systems, *Sov. J. Comput. Syst. Sci.,* 29(5), 126, 1991.

66. **Chernousko, F. L.,** Decomposition and suboptimal control in dynamic systems, *Optimal Control Appl. and Methods,* 14, 1993 in press.

67. **Chernousko, F. L., Akulenko, L. D., and Bolotnik, N. N.,** Time-optimal control for robotic manipulators, *Optimal Control Appl. Methods,* 10, 293, 1989.

68. **Chernousko, F. L., Akulenko, L. D., and Sokolov, B. N.,** *Control of Oscillations,* Nauka, Moscow, 1980 (in Russian).

69. **Chernousko, F. L., Bolotnik, N. N., and Gradetsky, V. G.,** *Manipulation Robots: Dynamics, Control, Optimization,* Nauka, Moscow, 1989.

70. **Chernousko, F. L. and Gradetsky, V. G.,** Dynamics of industrial robots with elastic flexible structure, in *Proc. 15th Int. Symp. on Industrial Robots,* Tokyo, 1985.

71. **Chernousko, F. L. and Lyubushin, A. A.,** Method of successive approximations for solution of optimal control problems, *Optimal Control Appl. Methods,* 3, 101, 1982.

72. **Chernousko, F. L., Pirumov, G. U., and Rogov, N. N.,** Modelling and control for manipulative robots with elastic elements, in *Proc. 8th CISM-IFtoMM Symp. on Theory and Practice of Robots and Manipulators,* RoManSy 8, Morecki, A., Bianchi, G., and Jaworek, K., Eds., Warszawa, 1992, 116.

73. **Chernousko, F. L. and Rogov, N. N.,** Optimal control of a robot with electric drive and elastic element, in *Robot Control 1988 (SYROCO' 88), Selected Papers from the 2nd IFAC Symposium,* Rembold, U., Ed., Pergamon Press, Oxford, 1989, 231.

74. **Collie, A. A., Billingsley, J., and von Puttkamer, E,** Design and performance of the Portsmouth climbing robot, in *Proc. 7th Int. Symp. on Automation and Robotics in Construction,* Vol. 1, Bristol, 1990, 16.

75. *Control Systems of Industrial Robots,* Makarov, I. M., Popov, E. P., and Chiganov, V. A., Eds., Mashinostroyeniye, Moscow, 1984 (in Russian).

76. **De Schutter, J., Van Brussel, H., Adams, M., Froment, A., and Faillot, J. L.,** Control of flexible robots using generalized non-linear decoupling, in *Robot Control 1988 (SYROCO' 88),* Rembold, U., Ed., VDI/VDE-GMA, Düsseldorf, 1988, 98.1.

77. **Desoyer, K., Kopacek, P., Lugner, P., and Troch, I.,** Flexible robots — a survey, in *Theory of Robots,* Kopacek, P., Troch, I., and Desoyer, K., Eds., Pergamon Press, Oxford, 1988, 23.

78. **Desoyer, K., Kopacek, P., Lugner, P., and Troch, I.,** Flexible robots, in *Intelligent Robotic Systems: Analysis, Design and Programming,* Tzafestas, S., Ed., Marcel Dekker, New York, 1988.

79. **Dimentberg, F. M.,** *Theory of Spatial Link Mechanisms,* Nauka, Moscow, 1982 (in Russian).

80. **Dmitriyev, V. N. and Gradetsky, V. G.,** *Fundamentals of Pneumatics,* Mashinostroyeniye, Moscow, 1973 (in Russian).

81. **Dubowsky, S. and Blubaugh, T. D.,** Time optimal robotic manipulator motions and work places for point to point tasks, in *Proc. 24th Conf. on Decision and Control,* Fort Lauderdale, 1985.

82. **Dubowsky, S., Norris, M. A., and Shiller, Z.,** Time optimal trajectory planning for robotic manipulators with obstacle avoidance: a CAD approach, in *Proc. IEEE Int. Conf. on Robotics and Automation,* San Francisco, 1986.

83. **Dubowsky, S., Norris, M. A., and Shiller, Z.,** Time optimal robotic manipulator task planning, in *Prepr. 6th CISM-IFToMM Symp. on Theory and Practice of Robots and Manipulators. Ro. Man. Sy.'86,* Crakow, 1986, 556.

84. **Dubowsky, S. and Shiller, Z.,** Optimal dynamic trajectories for robotic manipulators, in *Theory and Practice of Robots and Manipulators. Proc. of Ro. Man. Sy.'84: 5th CISM-IFToMM Symp.,* Morecki, A., Bianchi, G., and Kedzior, K., Eds., Kogan Page, London, 133.

85. **Eliseyev, S. V., Kuznetsov, N. K., and Lukyanov, A. V.,** *Control of Oscillations of Robots,* Nauka, Novosibirsk, 1990 (in Russian).

86. **Feshchenko, S. F., Shkil, N. I., and Nikolenko, L. D.,** *Asymptotic Methods in Theory of Linear Differential Equations,* Naukova Dumka, Kiev, 1982 (in Russian).

87. **Fu, K. S., Gonzalez, R. C., and Lee, C. S. G.,** *Robotics: Control, Sensing, Vision, and Intelligence,* McGraw-Hill, New York, 1987.

88. **Fujita, A., Tsuge, M., Mori, K., Sonoda, S., Watahiki, S., and Ozaki, N.,** Development of inspection robots for spherical gas storage tanks, in *Proc. 16th Int. Symp. on Industrial Robots,* Brussels, 1986, 1185.

89. **Fukuda, T.,** Control of flexible robotic arms, *Bull. JSME,* 29, 1269, 1986.

90. **Fukuda, T. and Kobayashi, H.,** Configuration control method of a control configured robot (CCR) consisting of multiple link-wheel mechanisms, in *Prepr. 10th World Congress on Automatic Control,* Vol. 4, Munich, 1987.

91. **Fukuda, T. and Kuribayashi, Y.,** Precise positioning control of flexible arms with reliable control system, in *Proc. of '83 Int. Conf. on Advanced Robotics,* Tokyo, 1983, 237.

92. **Geering, H. P., Guzzella, L., Hepner, S. A. R., and Onder, C. H.,** Time-optimal motions of robots in assembly tasks, *IEEE Trans. Autom. Control,* AC-31, 512, 1986.

93. **Geradin, M., Robert, G., and Bernardin, C.,** Dynamic modelling of manipulators with flexible members, in *Advanced Software in Robotics,* Danthine, A. and Geradin, M., Eds., Elsevier Science Publishers, 1984, 27.

94. **Gorbachev, N. N., Kim, D. P., and Shukhov, A. G.,** Synthesis of control algorithms on the basis of the inverse problem of dynamics with allowance for a constraint on the control, *Sov. J. Comput. Syst. Sci.,* 26(1), 142, 1988.

95. **Gorinevskii, D. M. and Formalskii, A. M.,** Stability of motion of an elastic manipulator with force feedback, *Mech. Solids (USSR),* 20(3), 48, 1985.

96. **Gradetsky, V. G.,** Primary transducers for robots using the properties of fluidic flows, in *Informatsionnyye i Upravlayushchiye Sistemy Robotov,* Institut Prikladnoi Matematiki An SSSR, Moscow, 1982 (in Russian).

97. **Gradetsky, V. G.,** Pneumatic measuring systems for industrial robots, *Vestn. Mashinostr.,* No. 7, 1982 (in Russian).

98. **Gradetsky, V. G.,** On the control of the gripping force in pneumatic grippers for industrial robots, in *Pneumatics and Hydraulics: Drives and Control Systems,* No. 11, Mashinostroyeniye, Moscow, 1984 (in Russian).

99. **Gradetsky, V. and Akselrod, B.,** Motion of wall surface robot under external disturbances, in *Proc. 20th Int. Symp. on Industrial Robots,* Tokyo, 1989, 261.

100. **Gradetsky, V. G., Akselrod, B. V., Gradetsky, A. V., and Dwivedi, S. N.,** Motion of climbing robot with manipulator under vibrations, in *Proc. Int. Robots and Vision Automation Conf.,* Detroit, 1991, 6–41.

101. **Gradetsky, V. G., Gukasyan, A. A., Grudev, A. I., and Chernousko, F. L.,** Effect of elastic structural compliance of robots on their dynamics, *Mech. Solids (USSR),* 20(3), 63, 1985.

102. **Gradetsky, V. G., Nazarov, V. V., Samvelyan, K. V., Stepanov, V. P., and Chernousko, F. L.,** Dynamics of an anthropomorphic robot involved into a robotic welding set up, *Izv. Akad. Nauk SSSR, Tekh. Kibern.,* 3, 50, 1987 (in Russian).

103. **Gradetsky, V. G. and Paroy, A. A.,** Industrial robot with smooth deceleration of the motion of pneumatic actuator, *Vestn. Mashinostr.,* No. 3, 1981 (in Russian).

104. **Gradetsky, V. G. and Rachkov, M. Yu.,** Pneumatic sensors for grippers of industrial robots, in *Pnevmoavtomatika,* Nauka, Moscow, 1982 (in Russian).

105. **Gradetsky, V. and Rachkov, M.,** Wall climbing robot and its applications for building construction, *Mech. Syst. Eng.,* 1, 225, 1990.

106. **Gradetsky, V. G., Rachkov, M. Yu., Sizov, Yu. G., Ulyanov, S. V., and Chernousko, F. L.,** Mobile systems with wall climbing robots, *Izv. Akad. Nauk SSSR, Tekh. Kibern.,* 6, 171, 1991 (in Russian).

107. **Gradetsky, V. G., Rachkov, M. Yu., Ulyanov, S. V., and Nandi, G. C.,** Robots for cleaning and decontamination of building construction, in *Proc. 8th Int. Symp. on Automation and Robotics in Construction,* Stuttgart, 1991, 257.

108. **Gradetsky, V. G., Rachkov, M. Yu., Ulyanov, S. V., Nandi, G. C., and Semenov, E. A.,** Mobile robotic complex for cleaning, *J. Flexible Manuf. Syst. Robotics,* 4, 21, 1991 (in Russian).

109. **Gradetsky, V. G., Ryzhov, N. N., Ermolayev, A. M., and Semin, V. V.,** Motion planning of technology robots in specific environments, in *Proc. 5th Int. Conf. on Advanced Robotics, 91 ICAR,* Vol. 2, Pisa, 1991, 1557.

110. **Gradetsky, V. G., Veshnikov, V. B., and Gukasyan, A. A.,** The influence of elastic properties of the industrial robot mechanism on the static positioning accuracy, in *Diagnosis of Equipment of Complex Automatic Manufacturing,* Nauka, Moscow, 1984 (in Russian).

111. **Gradetsky, V. G. and Yevdokimov, A. I.,** Analysis of the control algorithms for programmed robotic manipulators, in *Struynaya Tekhnika. Doklady VI Mezhdunarodnoy Konferentsii "Yablonna",* Nauka, Moscow, 1976 (in Russian).

112. **Hanafusa, H., Yoshikawa, T., Nakamura, Y., and Takoda, M.,** Contouring control of an articulated robot arm with manipulation variable feedback, in *Proc. 11th Int. Symp. on Industrial Robots,* Tokyo, 1981.

113. **Hanafusa, H., Yoshikawa, T., and Nakamura, Y.,** Analysis and control of articulated robot arm with redundancy, in *Prepr. of the 8th IFAC World Congress,* August 1981, XIV-78.

114. **Heimann, B., Loose, H., Schmidt, K. D., Rothe, H., and Lyubushin, A. A.,** Dynamics and optimal control of manipulation robots, *Adv. Mech.,* 7, 1984 (in Russian).

115. **Hemami, A.,** Studies on a light weight and flexible robot manipulator, *Robotics,* 1, 27, 1985.

116. **Hirose, S.,** Wall climbing vehicle using internally balanced magnetic unit, in *Prepr. 6th CISM-IFToMM Symp. on Theory and Practice of Robots and Manipulators, Ro. Man. Sy' 86,* Crakow, 1986, 363.

117. **Hohenbichler, G., Plöckinger, P., and Lugner, P.,** Comparison of a modal-expansion- and a finite-element-model for a two-beam flexible robot arm, in *Robot Control 1988 (SYROCO'88),* Rembold, U., Ed., VDI/VDE-GMA, Düsseldorf, 1988, 3.1.

118. **Hollerbach, J. M. and Suh, K. C.,** Redundancy resolution of manipulators through torque optimization, *IEEE J. Robotics Autom.,* RA-3, 308, 1987.

119. **Huang, H.-P. and McClamroch, N. H.,** Time-optimal control for a robotic contour following problem, *IEEE J. Robotics Autom.,* 4, 140, 1988.

120. *Information and Control Systems of Robots,* **Okhotsimskii, D. Ye.,** Ed., Institut Prikladnoy Matematiki AN SSSR, Moscow, 1982.

121. **Ishlinskii, A. Yu.,** *Orientation, Gyroscopes, and Inertial Navigation,* Nauka, Moscow, 1976 (in Russian).

122. **Ishlinskii, A. Yu., Chernousko, F. L., and Gradetsky, V. G.,** Some problems of mechanics and control for pneumatic industrial robots, in *Proc. 13th Int. Symp. on Industrial Robots,* v. 2, Chicago, 1983, 1369.

123. **Johanni, R.,** On the automatic generation of the equations of motion for robots with elastically deformable arms, in *Theory of Robots, IFAC Proc.,* Ser. 3, 1988, 143.

124. **Judd, R. P. and Falkenburg, D. R.,** Dynamics of non-rigid articulated robot linkages, *Trans. Autom. Control,* AC-30, 499, 1985.

125. **Kahn, M. E. and Roth, B.,** The near-minimum-time control of open-loop articulated kinematic chains, *Trans. ASME J. Syst. Meas. Control,* 93, 165, 1971.

126. **Kalra, P. and Sharan, A. M.,** Accurate modelling of flexible manipulators using finite element analysis, *Mech. Mach. Theory,* 26, 299, 1991.

127. **Kaplunov, A. A,** Optimization of motions of a kinematically redundant manipulator, *Izv. Akad. Nauk SSSR, Tekh. Kibern.,* 1, 158, 1984 (in Russian).

128. **Kärkkäinen, P.,** On the control of manipulator flexible motion by modal-space techniques, *Acta Polytech. Scand., Math. Comput. Sci. Ser.,* 14, 2, 1985.

129. **Kärkkäinen, P.,** Compensation manipulator flexibility effect by modal space techniques, in *IEEE Conf. of Robotics and Automation,* St. Louis, 1985.

130. **Kazerooni, H., Waibel, B. J., and Kim, S.,** On the stability of robot compliant motion control: theory and experiments, *Trans. ASME J. Dyn. Syst. Meas. Control,* 112, 417, 1990.

131. **Kazerounian, K.,** Optimal manipulation of redundant robots, *Int. J. Robotics Autom.,* 2, 54, 1987.

132. **Kazerounian, K. and Wang, Z.,** Global versus local optimization in redundancy resolution of robotic manipulator, *Int. J. Robotics Res.,* 7, 3, 1988.

133. **Khadem, S. E. and Dubey, R. V.,** A global Cartesian space obstacle avoidance scheme for redundant manipulators, *Optimal Control Appl. Methods,* 12, 279, 1991.

134. **Khoukhi, A. and Hamam, Y.,** Optimal time-energy trajectory planning for robots with hard constraints, *Control Theory Adv. Technol.,* 6, 417, 1990.

135. **Kiriazov, P. and Marinov, P.,** Control synthesis of manipulator dynamics in handling operations, *Teor. Prilozhna Mekh.,* 2, 15, 1983.

136. **Kiriazov, P. and Marinov, P.,** Robot control synthesis in conjunction with moving workpieces, in *Prepr. 6th CISM-IFToMM Symp. on Theory and Practice of Robots and Manipulators, Ro. Man. Sy.'86,* Crakow, 1986, 284.

137. **Klimov, D. M., Gradetsky, V. G., and Teryayev, E. D.,** Some problems of design and application of industrial robots and robotic systems, in *Proc. SICIR-87,* Tokyo, 1987, 289.

138. **Kobrinskii, A. A. and Kobrinskii, A. E.,** *Manipulation Systems of Robots,* Nauka, Moscow, 1984 (in Russian).

139. **Konzelmann, J., Bock, H. G., and Longman, R. W.,** Time-optimal extension or retraction in polar coordinate robots: a numerical analysis of the switching structure, in *Proc. of the AIAA Guidance, Navigation, and Control Conference,* Boston, August, 1989, 883.

140. **Konzelmann, J., Bock, H. G., and Longman, R. W.,** Time-optimal trajectories of elbow robots by direct methods , in *Proc. of the AIAA Guidance, Navigation, and Control Conference,* Boston, August, 1989, 895.

141. **Konzelmann, J., Bock, H. G., and Longman, R. W.,** Time-optimal trajectories of polar robot manipulators by direct methods, *Modeling Simulation,* 20, 1933, 1989.

142. **Kopacek, P., Desoyer, K., and Lugner, P.,** Modelling of flexible robots — an introduction, in *Robot Control 1988 (SYROCO'88),* Rembold, U., Ed., VDI/VDE-GMA, Düsseldorf, 1988, 1.1.

143. **Korytko, O. B. and Yudin, V. I.,** On calculation of eigenfrequencies for a manipulator of an industrial robot in the general case, in *Control of Robotic Systems and their Sensorization,* Nauka, Moscow, 1983 (in Russian).

144. **Kozlov, V. V., Makarychev, V. P., Timofeyev, A. V., and Yurevich, E. I.,** *Dynamics of Robot Control,* Nauka, Moscow, 1984 (in Russian).

145. **Kozyrev, Yu. G.,** *Manipulation Robots: a Handbook,* Mashinostroyeniye, Moscow, 1983 (in Russian).

146. **Krutko, P. D.,** *Inverse Problems of Dynamics of Control Systems,* Nauka, Moscow, 1988 (in Russian).

147. **Krutko, P. D.,** *Control of Robot Actuators,* Nauka, Moscow, 1991 (in Russian).

148. **Lakota, N. A. and Rakhmanov, E. V.,** Finite-element method in dynamics of an elastic manipulator, *Izvestiya vuzov, Mashinostr.,* 5, 51, 1985 (in Russian).

149. **Lakota, N. A., Rakhamanov, E. V., and Shvedov, V. N.,** Control of an elastic manipulator on a trajectory, *Izv. Akad. Nauk SSSR, Tekh. Kibern.,* 2, 53, 1980 (in Russian).

150. **Lavrovskii, E. K. and Formalskii, A. M.,** Stabilization of a given position of an elastic rod, *J. Appl. Math. Mech. (USSR),* 53, 590, 1989.

151. **Lavrovskii, E. K. and Formalskii, A. M.,** On the stabilization of the angular position of an elastic beam, *Izv. Akad. Nauk SSSR, Tekh. Kibern.,* 6, 115, 1989 (in Russian).

152. **Lavrovskii, E. K. and Formalskii, A. M.,** Stabilization of the control process of a bar in the presence of longitudinal or torsional deformations, *Izv. Akad. Nauk SSSR, Tekh. Kibern.,* 6, 203, 1991 (in Russian).

153. **Liegeois, A.,** Automatic supervisory control of the configuration and behaviour of multibody mechanisms, *IEEE Trans. Syst. Man Cybern.,* SMC-7, 868, 1977.

154. **Lilov, L. and Wittenburg, J.,** Dynamics of chains of rigid bodies and elastic rods with revolute and prismatic joints, in *IUTAM/IFToMM Symp. "Dynamics of Multibody Systems",* Bianchi, G. and Schiehlen, W., Eds., Springer-Verlag, Berlin, 1985, 141.

155. **Long, T. W. and Carrington, C. K.**, A new non-linear optimal control method: the tracking and relative direction imbedded system, *Int. J. Control*, 54, 417, 1991.

156. **Loose, H., Rothe, H., and Schmidt, C.-D.**, Verfahren und Programme zur Optimalen Steuerung von Industrierobotern, *Z. Angew. Math. Mech.*, 64, M 476, 1984.

157. **Luh, J. Y. S. and Lin, C. S.**, Optimum path planning for mechanical manipulators, *Trans. ASME J. Dyn. Syst. Meas. Control*, 102, 142, 1981.

158. **Lurye, A. I.**, *Analytical Mechanics*, Fizmatgiz, Moscow, 1961 (in Russian).

159. **Maizza-Netto, O.**, Modal approach for modelling flexible manipulators, in *Automatic Control in Space: Proc. of the 8th IFAC Symposium*, Pergamon Press, Oxford, 1980, 405.

160. **Marinov, P. and Kiriazov, P.**, Synthesis of time-optimal control for manipulator dynamics, *Teor. Prilozhna Mekh.*, 1, 13, 1984.

161. **Marinov, P. and Kiriazov, P.**, A direct method for optimal control synthesis of manipulator point-to-point motion, in *Prepr. 9th IFAC World Congress, Budapest, Hungary, 1984*, Vol. IX, MacFarlane, A. G. J. and Rauch, H. E., Eds., Budapest, 1984, 219.

162. **Matsuoka, K. and Citron, S. J.**, Symbolic processing and dynamic simulation of equations of motion for flexible manipulators, in *Proc. ISA/85*, 1985, 1601.

163. **Meier, E.-B. and Bryson, A. E.**, Efficient algorithm for time-optimal control of a two-link manipulator, *J. Guidance*, 13, 859, 1990.

164. **Midha, A., Erdman, A G., and Frorib, D. A.**, Finite element approach to mathmatical modelling of high-speed elastic linkages, *Mech. Mach. Theory*, 13, 603, 1978.

165. **Mikhailov, S. A.**, Natural oscillations of a two-element link with a point mass, *Mech. Solids (USSR)*, 18(2), 66, 1983.

166. **Mikhailov, S. A. and Chernousko, F. L.**, Investigation of the dynamics of a manipulator with elastic elements, *Mech. Solids (USSR)*, 19(2), 49, 1984.

167. **Mikhailov, S. A. and Chernousko, F. L.**, Dynamics of an elastic manipulator with specified control moments or motions of the load, *Mech. Solids (USSR)*, 19(5), 17, 1984.

168. **Mikhailov, S. A. and Satovskaya, O. L.**, Elastic vibrations of an anthropomorphic manipulator, *Mech. Solids (USSR)*, 24(1), 58, 1989.

169. **Mitropolskii, Yu. A.**, *Problems of Asymptotic Theory of Nonstationary Oscillations*, Nauka, Moscow, 1964 (in Russian).

170. **Nakamura, Y. and Hanafusa, H.**, Optimal redundancy control of robot manipulators, *Int. J. Robotics Res.*, 6, 32, 1986.

171. **Nayfeh, A. H.**, *Perturbation Methods*, John Wiley & Sons, New York, 1973.

172. **Nevins, J., Desai, M., Fogel, E., Walker, B., and Whitney, D.**, Adaptive control, learning and cost effective sensor systems for robotic or advanced automation systems, in *Proc. of 83 ICAR*, Tokyo, 1983.

173. **Newman, W. S.**, Time optimal control of balanced manipulators, *Trans. ASME J. Dyn. Syst. Meas. Control*, 111, 187, 1989.

174. **Nicosia, S., Nicoló, F., and Lentini, D.**, Dynamic control of industrial robots with elastic and dissipative joints, in *Proc. 8th Triennial IFAC World Congress*, Kyoto, 1981, 1933.

175. **Nicosia, S., Tomei, P., and Tornambe, A.**, Dynamic modelling of flexible robot manipulators, in *Proc. IEEE Conf. Robotics Autom.*, 1, 1986, 365.

176. **Nishi, A. and Miyagi, H.**, Control of a wall-climbing robot using propulsive force of propeller, in *Proc. IEEE/RS Int. Workshop Intel. Robotics and Syst. 91*, Vol. 3, Tokyo, 1991.

177. **Nitta, Y.**, Visual identification and sorting with TV-camera applied to automated inspection apparatus, in *Proc. 10th Int. Symp. on Industrial Robots*, Milan, 1980.

178. **Nof, S. Y.**, *Handbook of Industrial Robotics*, IFS Publications, 1986.

179. **Nosov, V. N., Troitskiy, A. V., and Troitskiy, V. A.**, Optimality problems in manipulator kinematics, *Sov. J. Comput. Syst. Sci.*, 26(6), 107, 1988.

180. **Nosov, V. N., Troitskii, A. V., and Troitskii, V. A.,** Some problems of optimization of the programmed motion of single-element manipulators, *Mech. Solids (USSR)*, 24(2), 33, 1989.

181. **Oberle, H.-J.,** Numerical computation of singular control functions for a two-link robot arm, in *Proc. of the Conference on Optimal Control and Variational Calculus*, Bulirsch, R., Miele, A., Stoer, J., and Well, K. H., Eds., Springer-Verlag, Berlin, 1987, 244.

182. **Okhotsimskii, D. Ye. and Golubev, Yu, F.,** *Mechanics and Control of the Motion of Automatic Walking Machine*, Nauka, Moscow, 1984 (in Russian).

183. **Okina, N., Minowa, K., and Okamoto, K.,** Three-dimensional measurement system by servomechanism, in *Proc. 11th Int. Symp. on Industrial Robots*, Tokyo, 1981.

184. **Osipov, S. N. and Formalskii, A. M.,** The problem of the time-optimal turning of a manipulator, *J. Appl. Math. Mech. (USSR)*, 52, 725, 1988.

185. **Pars, L. A.,** *A Treatise on Analytical Dynamics*, Oxbow Press, Woodbridge, CT, 1979.

186. **Paul, R. P.,** *Robot Manipulator: Mathematics, Programming and Control*, MIT Press, Cambridge, MA, 1981.

187. **Petrov, B. A.,** *Manipulators*, Mashinostroyeniye, Leningrad, 1984 (in Russian).

188. **Pfeiffer, F.,** Geometrical solution of a manipulator optimization problem, in *Control Applications of Nonlinear Programming and Optimization 1989, Proc. of the 8th IFAC Workshop*, Siguerdidjane, H. B. and Bernhard, P., Eds., Pergamon Press, Oxford, 1989, 83.

189. **Pfeiffer, F., Gebler, B., and Kleeman, U.,** On dynamics and control of elastic robots, in *Robot Control 1988 (SYROCO'88)*, Rembold, U., Ed., Pergamon Press, Oxford, 1988.

190. **Pfeiffer, F. and Johanni, R.,** A concept for manipulator trajectory planning, *IEEE J. Robotics Autom.*, RA-3, 115, 1987.

191. **Pirumov, G. U.,** Parametric optimization of an acceleration-deceleration arm of manipulator, *Sov. J. Compt. Syst. Sci.*, 29(1), 101, 1991.

192. **Pontryagin, L. S., Boltyanski, V. G., Gamkrelidze, R. V., and Mishchenko, E. F.,** *Mathematical Theory of Optimal Processes*, Wiley-Interscience, New York, 1962.

193. **Popov, E. P., Vereshchchagin, A. F., and Zenkevich, S. L.,** *Manipulation Robots: Dynamics and Algorithms*, Nauka, Moscow, 1978, (in Russian).

194. **Rajan, V. T.,** Minimum time trajectory planning, in *Proc. of the 1985 IEEE Conf. on Robotics and Automation*, St. Louis, MO, 1985, 759.

195. **Rakhamanov, E. V., Strelkov, A. N., and Shvedov, B. N.,** Development of mathematical model for an elastic manipulator on a moving base, *Izv. Akad. Nauk SSSR, Tekh. Kibern.*, 4, 109, 1981 (in Russian).

196. **Rauh, J. and Schiehlen, W.,** A unified approach for the modelling of flexible robot arms, in *Prepr. of 6th CISM-IFToMM Symp. on Theory and Practice of Robots and Manipulators, Ro. Man. Sy' 86*, Crakow, 1986, 74.

197. **Rogov, N. N.,** On the deceleration of a two-mass vibrating system, *Sov. J. Comput. Syst. Sci.*, 26(6), 107, 1988.

198. **Rogov, N. N. and Chernousko, F. L.,** Optimal control of the electric motor of a robot with an elastic element, *Sov. J. Comput. Syst. Sci.*, 27(5), 39, 1989.

199. **Sahar, G. and Hollerbach, J. M.,** Planning of minimum-time trajectories for robot arms, *Int. J. Robotics Res.*, 5, 90, 1986.

200. **Sakawa, Y. and Matsuno, F.,** Modelling and control of a flexible manipulator with a parallel drive mechanism, *Int. J. Control*, 44, 299, 1986.

201. **Sakawa, Y., Matsuno, F., and Fukushima, S.,** Modelling and control of a flexible robot arm, *J. Robotic Syst.*, 2, 453, 1985.

202. **Sato, O., Shimojima, H., and Kitamura, Y.,** Minimum-time control of a manipulator with two degrees of freedom, *Bull. JSME*, 26, 1404, 1983.

203. **Sato, O., Shimojima, H., Kitamura, Y., and Yoinara, H.,** Minimum-time control of a manipulator with two degrees of freedom (2nd Report, Dynamic characteristics of gear train and axes), *Bull. JSME*, 28, 959. 1985.

204. **Sato, O., Shimojima, H., and Yoinara, H.,** Minimum-energy control of a manipulator with two degrees of freedom, *Bull. JSME,* 29, 573, 1986.

205. **Satovskaya, O. L.,** Dynamic of a manipulator with an elastic extended load, *Izv. Akad. Nauk SSSR, Tekh. Kibern.,* 3, 233, 1991 (in Russian).

206. **Sattar, T. P. and Bolotnik, N. N.,** Adaptive time-optimal control of a manipulation robot, Izv. Akad. Nauk, *Tekh. Kibern.,* 3, 171, 1993.

207. **Schmitt, D., Soni, A. H., Srinvasan, V., and Nagamathan, G.,** Optimal motion programming of robot manipulators, *J. Mech. Transmissions Autom. Design,* 107, 239, 1985.

208. *Sensory Systems and Adaptive Industrial Robots,* Popov, E. P. and Kluyev, V. V., Eds., Mashinostroyeniye, Moscow, 1985 (in Russian).

209. **Shiller, Z. and Lu, H.-H.,** Computation of path constrained time optimal motions with dynamic singularities, *Prepr. of the Laboratory for Robotics Automation and Manufacturing,* Department of Mechanical, Aerospace and Nuclear Engineering, University of California, Los Angeles, January 1991.

210. **Singh, T., Dubey, R. N., and Golnaraghi, M. F.,** Effects of joint flexibility on the motion of a flexible-arm robot, *Dyn. Stability Syst.,* 6, 235, 1991.

211. **Skaar, S. B. and Tucker, D.,** Point control of a one-link flexible manipulator, *J. Appl. Mech.,* 53, 23, 1986.

212. **Sliede, P. B., Auzinsh, Ya. P., and Itkin, V. M.,** Algorithms for mathematical simulation of elastic manipulators using a computer, in *Symposium Grundlagen der Dynamik und Steuerung von Industrierobotern, Vorträge, Band 1,* Berlin, 1985.

213. **Sliede, P. B., Itkin, V. M., and Auzinsh, Ya. P.,** Algorithm for calculation of oscillatory characteristics of manipulation robots using a computer, *Mashinovedeniye,* 2, 48, 1984 (in Russian).

214. **Smolnikov, B. A.,** *Problems of Mechanics and Optimization of Robots,* Nauka, Moscow, 1991 (in Russian).

215. **Sontag, E. D. and Sussmann, H. J.,** Time-optimal controls of manipulators, in *Proc. IEEE Conference on Robotics and Automation,* Philadelphia, 1988, 370.

216. **Soudunsaari, R., Gradetsky, V., Veshnikov, V., and Abarinov, A.,** Active pneumatic remote center compliance system, in *Proc. 20th Int. Symp. on Industrial Robots,* Tokyo, 1989, 717.

217. **Steinbach, M., Bock, H.-G., and Longman, R.,** Time optimal control of SCARA robots, in *Proc. of the AIAA Guidance, Navigation, and Control Conference,* Portland, OR, 1990.

218. **Stepanenko, Yu. A.,** Some problems of optimal control of manipulators, *Mekh. Mash.,* 22, 86, 1969 (in Russian).

219. **Stepanenko, Yu. A.,** Problem of optimal control of a manipulator, in *Theory of Machines of Automatic Action,* Nauka, Moscow, 1970 (in Russian).

220. **Sujiyata, S., Naiton, S., Sato, C., Ozaki, N., and Watahiki, S.,** Wall surface vehicles with magnetic legs or vacuum legs, in *Proc. 16th Int. Symp. on Industrial Robots,* Brussels, 1986, 691.

221. **Sunada, W. and Dubowsky, S.,** The application of finite element methods to the dynamic analysis of flexible spatial and co-planar linkage systems, *J Mech. Design,* 103, 643, 1981.

222. **Sunada, W. and Dubowsky, S.,** On the dynamic analysis and behavior of industrial robotic manipulators with elastic elements, *Trans. ASME, J Mech. Transmissions Autom. Design,* 105, 42, 1983.

223. **Takano, M. and Susaki, K.,** Time optimal control of PTP motion of a robot with collision avoidance, in *Proc. 3rd Conf. on Robotics, ICAR 87,* Versailles, 1987.

224. **Timofeev, A. V.,** Adaptive control of robots, *Sov. J. Comput. Syst. Sci.,* 27(5), 59, 1989.

225. **Tosunogly, S., Lin, S.-H., and Tesar, D.**, Complete accessibility of oscillations in robotic systems by orthononal projections, *Trans. ASME. J. Dyn. Syst. Meas. Control,* 112, 194, 1990.

226. **Troch, I.**, Time-optimal path generation for continuous and quasi-continuous path control of industrial robots, *J. Intelligent Robotic Syst.,* 2, 1, 1989.

227. **Troch, I.**, Time-suboptimal quasi-continuous path generation for industrial robots, *Robotica,* 7, 297, 1989.

228. **Troch, I. and Kopacek, P.**, Control concepts and algorithms for flexible robots — an expository survey, in *Robot Control 1988 (SYROCO'88),* Rembold, U., Ed., VDI/VDE-GMA, Düsseldorf, 1988, 2.1.

229. **Truckenbrodt, A.**, Regelung eines Flexiblen Manipulators, *ZAMM,* 58, 184, 1978.

230. **Truckenbrodt, A.**, Dynamics and control methods for moving flexible structures and their application to industrial robots, in *Proc. 5th World Congress on Theory of Machines and Mechanisms,* ASME, 1979.

231. **Truckenbrodt, A.**, Bewegungsverhalten und Regelung Hybrider Mekrhörpersysteme mit Anwendung auf Industrieroboter, in *VDI-Bericht,* 33, Reihe 8, 1890.

232. **Truckenbrodt, A.**, Zur Vernachlässigung Höherfrequenter Schwingungsformen bei Mechanischen Systemen, *ZAMM,* 61, 65, 1981.

233. **Truckenbrodt, A.**, Modelling and control of flexible manipulator structures, in *Proc. 4th CISM-IFToMM Symposium on the Theory and Practice of Robots and Manipulators,* Zabarów, 1981, 90.

234. **Truckenbrodt, A.**, Truncation problems in the dynamics and control of flexible mechanical systems, in *Proc. 8th Triennial IFAC World Congress,* Kyoto, 1981.

235. **Truckenbrodt, A.**, Regelung elastischer mechanischer Systeme, *Regelungstechnik,* 30, 277, 1982.

236. **Usoro, P. B., Nadira, R., and Mahil, S. S.**, A finite element/Lagrange approach to modelling lightweight flexible manipulators, *Trans. ASME.,* 108, 198, 1986.

237. **Van Brussel, H.**, Sensor based robots do work, in *Proc. 21st Int. Symp. on Industrial Robots,* Copenhagen, 1990, 5.

238. **Vasilyeva, A. B. and Butuzov, V. F.**, *Asymptotic Expansions for Solutions of Singularly Perturbed Systems,* Nauka, Moscow, 1973 (in Russian).

239. **Vorobyov, E. I.**, Influence of bending stiffness of a robot arm on its extension under bang-bang control, in *Dinamika Mashin,* No. 51, Nauka, Moscow, 1976 (in Russian).

240. **Vorobyov, E. I. and Shchogoleva, A. N.**, Minimum-time optimization of pneumatic manipulator by choosing the switch instants for drives, *Mashinovedeniye,* 3, 24, 1978.

241. **Vukobratovic, M.**, *Applied Dynamics of Manipulation Robots,* Springer-Verlag, Berlin, 1989.

242. **Vukobratovic, M. and Kircanski, M.**, A method for optimal synthesis of manipulation robotic trajectories, *Trans. ASME. J. Dyn. Syst. Meas. Control,* 102, 69, 1980.

243. **Vukobratovic, M. and Kircanski, M.**, A dynamic approach to nominal trajectory synthesis for redundant manipulators, *IEEE Trans. Syst. Man Cybern.,* SMC-14, 1984.

244. **Vukobratovic, M. and Kircanski, M.**, *Kinematics and Trajectory Synthesis of Manipulation Robots,* Springer-Verlag, Berlin, 1985.

245. **Vukobratovic, M. and Potkonjak, V.**, Numerical method for simulating the dynamics of a manipulator with elastic properties, *Izv. Akad. Nauk SSSR, Tekh. Kibern.,* 5, 131, 1981 (in Russian).

246. **Vukobratovic, M. and Potkonjak, V.**, *Dynamics of Manipulation Robots,* Springer-Verlag, Berlin, 1982.

247. **Vukobratovic, M. and Stokic, D.**, *Applied Control of Manipulation Robots,* Springer-Verlag, Berlin, 1989.

248. **Walendy, U. and Weber, M.**, Ein Regelalgorithmus zum Schwingungsfreien Positionieren für einen weichen Manipulator bei einer r, φ — Bewegung, in *Symp. "Grundlagen der Dynamik und Steuerung von Industrierobotern",* Vorträge, Band 1, Berlin, 1985.

249. **Weinreb, A. and Bryson, A. E.**, Optimal control systems with hard control bounds, *IEEE J. Autom. Control*, AC-30, 1135, 1985.
250. **Weinreb, A. and Bryson, A. E.**, Minimum-time control of a two-link robot arm, in *Proc. Fifth IFAC Workshop on Nonlinear Programming and Optimization*, Capri, 1985.
251. **Whitney, D. E.**, Resolved motion rate control of manipulators and human prostheses, *IEEE Trans. Man-Machine Syst.*, MMS-10, 47, 1969.
252. **Whitney, D. E.**, The mathematics of coordinated control of prosthetic arms and manipulators, *Trans. ASME. J. Dyn. Syst. Meas. Control*, 94, 303, 1972.
253. **Wie, B., Chuang, C.-H., and Sunkel, J.**, Minimum-time pointing control of a two-link manipulator, *J. Guidance*, 13, 867, 1990.
254. **Wittenburg, J.**, *Dynamics of Systems of Rigid Bodies*, B. G. Teubner, Stuttgart, 1977.
255. **Yashi, O. S. and Ozgoren, K.**, Minimal joint motion optimization of manipulator with extra degrees of freedom, *Mech. Mach. Theory*, 19, 325, 1984.
256. **Yeung, K. S. and Chen, Y. P.**, Sliding-mode controller design of a single-link flexible manipulator under gravity, *Int. J. Control*, 52, 101, 1990.
257. **Yudin, V. I.**, Analysis of oscillations of a manipulator arm, *Prikl. Mekh.*, 16(10), 108, 1980 (in Russian).
258. **Yurevich, E. I., Avetikov, B. G., Korytko, O. B., Andrianov, Yu. D., Korolyov, B. A., and Savin, V. G.**, *Structure of Industrial Robots*, Mashinostroyeniye, Leningrad, 1980 (in Russian).
259. **Zak, V. L. and Pirumov, G. U.**, Modelling of the dynamics of an elastic manipulator with electromechanical drives, *Sov. J. Comput. Syst. Sci.*, 26(2), 39, 1988.
260. **Zak, V. L. and Pirumov, G. U.**, Optimal exponential law of acceleration-deceleration of a manipulator, *Sov. J. Comput. Syst. Sci.*, 26(6), 123, 1988.
261. **Zaremba, A. T.**, Dynamic model of plane elastic manipulator, *Mech. Solids (USSR)*, 20(5), 20, 1985.
262. **Zhang, W. and Wang, R. K. C.**, Collision-free time optimal control of a two-link manipulator, *Int. J. Robotics Autom.*, 1, 96, 1986.

INDEX